"Rivers and the inhabitants of the watery elements
are made for [the wise] to contemplate and for fools
to pass by without consideration."

Adapted from **Izaak Walton**, *The Compleat Angler* (1653)

"How pleasant the banks of the clear-winding Devon,
With green-spreading bushes, and flow'rs blooming fair!"

Robert Burns, *'The Banks of the Devon'* (1787)

THUNDER

A silence fills the soul
When thunders roll
Along the rigid hills;
And now the hills are fled.
Ghastly as dying eyes
The daylight dies
Within the cloud-wrapt peaks.
Wildly the shouts are sped,
And silence falls.
Again the thunder calls
As waterfalls
Pounding down sombre steeps.
Dreadful the forest cowers
Beneath the hissing rain:
Spirits in pain
Swish fiercely through the leaves,
Tearing the green-hung bowers
Ere silence falls.
Distant, more distant still,
Beyond the hill,
The voices die away.
Godlike the mountains dim
Rise from the welt'ring haze.
Trees gladly raise
Their dripping arms in joy.
Again a sky-lark's hymn
On silence falls.

William Soutar (1923)

'If Rivers Could Sing'

A Scottish River Wildlife Journey

A Year in the Life of the River Devon
as it flows through the Counties of
Perthshire, Kinross-shire & Clackmannanshire

Keith Broomfield

TIPPERMUIR
· BOOKS LIMITED ·

This first edition published and copyright 2020 by
Tippermuir Books Ltd, Perth, Scotland.
mail@tippermuirbooks.co.uk ~ www.tippermuirbooks.co.uk

ISBN 978-1-913836-00-9 (paperback)
A CIP catalogue record for this book is available from the British Library.

Project Coordination by Dr Paul S Philippou.
Illustrations/Maps: Rob Hands.
Cover Design by Matthew Mackie.
Editorial Support: Ajay Close, Jean Hands, Alan Laing, and Steve Zajda.
Text Design, Layout, and Artwork by Bernard Chandler [graffik].
Text set in Dante MT Std 10.5/14pt.

Printed and Bound by Ashford Colour Press, Gosport.

To my wonderful wife, Lynda

ACKNOWLEDGEMENTS

Writing a first book is always a daunting, yet satisfying, endeavour, and I would like to thank Paul Philippou of Tippermuir Books for embracing my original idea for this book and being so supportive throughout the whole writing and production process. Rob Hands also deserves special praise for his wonderful illustrations. A debt of thanks, go to Jean Hands, Alan Laing, and Steve Zajda for their readings and checking of earlier versions of this book, and to Ajay Close for her answers to editorial queries.

Special thanks, too, go to Janet Peck for volunteering so readily to take me on a canoe trip down the River Devon. The support of the Devon Angling Association in providing insight and knowledge of the river was tremendously useful, and I would like to thank Bryan Anderson, David Mudie, Alan Graham, Alan Armstrong and Colin Smail. Thanks also to Alison Brooks-Baker and Jo Girvan of the Forth Rivers Trust. *While writing this book, changing 'Coronavirus Lockdown' restrictions were imposed in Scotland. These were complied with fully and visits to the Devon were undertaken within the restrictions.*

CONTENTS

THE COURSE of the RIVER DEVON

The River Devon originates at
about 1,800 feet above sea level on Alva Moss
in the Ochil Hills in Central Scotland.

Approximately 33 miles long, the
Devon flows in a U-shaped pattern,
heading eastwards where it supplies
the water for Upper and Lower
Glendevon Reservoirs, then south-
east through Glendevon, where it
has been dammed near Muckhart
to create Castlehill Reservoir.

From there, it continues south-
eastwards, before abruptly
changing to a westerly course
at Crook of Devon, through
the gorges and waterfalls at
Rumbling Bridge and
Cauldron Linn.

It then it runs parallel to the
southern scarp of the Ochils
past Dollar, Tillicoultry, Alva
and Menstrie, and finally southwards on its last leg towards its
estuary on the Inner Forth
by Cambus near Alloa.

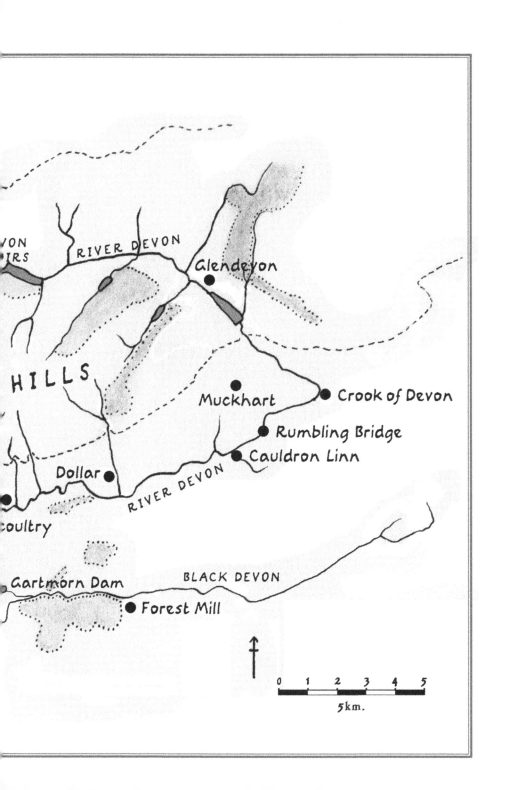

PROLOGUE
(June)

If rivers could sing, then this would be their song, a gentle gurgling chorus set against the rustle of wind caressed alders – a place of peace and tranquillity where the mind empties and nature prevails. I am sitting on a shingle bank in late June 2019 by the edge of the River Devon (hereafter the Devon) in Clackmannanshire. The sun is shining and by my feet azure-petalled water forget-me-nots sway in the breeze and a dipper, its white breast glinting in the light, bobs on a rock in the middle of the water's flow.

This is my river, a place I've come to know well over recent years and somewhere that has become part of my being. It is a place that stirs the emotions and brings back eclectic memories: a pair of otters playing by a far bank; a streaking flash of electric-blue as a kingfisher whirrs upriver; and the joy of hooking my first ever salmon on a river fly – and the despair of losing the same fish after a desperate battle.

Water Forget-me-not

Another time when angling, I stumbled in the water by a steep sloping shelf, my waders rapidly filling and the pull of the current dragging me under. The river was no longer a benign friend, but a brutal enemy, so much stronger than a mortal human. I floundered and flopped, but made my way back to the bankside, gasping, cold and a little bit scared. Respect the river, I thought to myself, always respect it, because if you stop doing so, then it will not respect you.

In summer the river's flow can be slow and gentle, the low water revealing tumbled tree trunks, many of which themselves must have been pulled under during times of spate, as happens periodically when the rain falls for any length of time. Then, the water rises, and the river spills over its banks, spreading mud-rippled waters like a creeping blanket across the

1

haugh and transforming the floor of the strath into an expansive sea, broken only by ribbons of higher ground and lone-sentinel trees.

Such thoughts breezed through my consciousness as I sat by the Devon's flow that summer afternoon, reflecting upon my passion for this river. A conundrum suddenly drifted across the mind. How could one fall in love with a river, a mere geographical feature, as I have so obviously done head-over-heels with the Devon? But I quickly turned the question on its head: how could you *not* fall in love with a river? For me, there is nothing more compelling than these ribbons of life that meander through our fields and hills; some rushing and urgent, others slow and tranquil. But whatever their nature, they are always places full of natural wonders, ranging from tiny bugs and microscopic algae to much larger creatures like the salmon and otter.

It is a river's power that really inspires: the tumbling waterfalls and bowl-shaped cauldrons forged by the forces of water over the millennia and the surging rapids and the crumbling banks, the ability to inundate surrounding fields in autumnal floods. Rivers never stand still, new channels develop, and shingle islands miraculously materialise, whilst others are destroyed and swept away. The river is as alive as any creature that lives within its bounds.

My earliest recollection of a river was as a four or five-year-old wandering down to the banks of the Water of Leith on the southern outskirts of Edinburgh accompanied by my older brother. I remember the green lushness and finding a dunnock's nest hidden in a moss-covered tree stump by the river's side, the four eggs shining like azure jewels. I recall the dampness of the air and the sound of the water bubbling over the rocks.

When the family then moved to the north of Edinburgh, the Water of Leith was still an ever-present friend, although slower in flow as it closed in upon the sea. I often watched water voles from the old disused railway bridge, hearing their 'plops' as they dived into the water and seeing their V-shaped wakes as they hurriedly swam for cover. From the same bridge I observed speckled brown trout, their heads facing upstream and tails moving rhythmically to keep pace with the current. I saw my first ever kingfisher here and watched tiny thread-like elvers making their way upstream after having travelled thousands of miles across the Atlantic from the Sargasso Sea. The spell had been cast.

In my early teens, and accompanied by friends, I ventured further afield to explore the River Tyne in East Lothian in a magnificent wooded valley near Humbie. Here, all caution was thrown to the wind and we would wade in the river exploring nooks and crannies and searching under banks and over-hangs for dipper nests. These intricate dome-shaped creations were easy to find once you knew where to look. Wading brought a new dimension to my connection with the river – it was like becoming an integral part of it – and you saw so much more.

Later, when I moved north to Aberdeen, the Water of Feugh – a tributary of the mighty River Dee near Banchory – became my favoured watercourse, a place to explore and enjoy. Here I would find the nests of grey wagtails clinging to rock ledges and watch woodcock at dusk engaging in their mystical territorial courtship flights known as 'roding'. I discovered my first slow worms basking in the sun only yards from the river's edge. The sandpipers were the stars in spring and early summer, flying on flickering wings from rock to rock accompanied by their high-pitched peeps. On boulders by the bank, it was not unusual to stumble upon the sweet-smelling droppings or spraints of otters.

On a different tributary of the Dee by the edge of the Cairngorms, there was a small gorge where rowan and birch trees miraculously found tenure on its steep rocky sides. In this dark and mysterious place, the river was narrow and only a few feet wide. The sides of this rocky cleft were home to peculiarly formed liverworts, primitive plants that thrive in the perpetual humidity and shade. But in late summer and early autumn the gorge held salmon in its dark watery depths. These fish had run up this tributary as far as they could go because the water was too low for them to negotiate the final leg over a spewing waterfall that provided access to the gravelly spawning grounds further upstream.

These salmon would lie in the deep pool in the gorge for weeks, waiting for the heavy rains that would deliver the right conditions to pass this seemingly impenetrable obstacle. One year, I stripped naked and donned a mask and snorkel and swam amongst the fish. The peaty water was freezing, and I could endure the cold for only so long, but the sight of their flashing torpedo-shaped bodies is still imprinted clearly upon my mind.

Then, there are the rivers I have come to know in other parts of the world, such as the Zambezi and Okavango in Zimbabwe and Botswana –

quite unlike any river at home, with their hippos and crocodiles, skimmers and tiger fish. Despite these differences, they too still maintain the same hypnotic attraction as any other river. More recently, and now with my young family in tow, the River Soča in Slovenia became a great favourite with its impossibly clear turquoise waters holding grayling and fine marbled and brown trout. We swam, rafted and canoed in the river and returned on three separate occasions such was its magnetic attraction.

River memories, oh yes, there are so many. But as I pondered such thoughts on this shingle shelf by the Devon on that tranquil summer's day, a dawning swept across me, another question arising in the mind. How well do I know my river, the Devon? What makes it tick, the wildlife that lives there – both obvious and those creatures and plants that are less so, and how do its forces sculpt the landscape? But it is more than just the nature, there is the impact on the river and its tributaries by humankind, too, both today and in the past when this river played its own little part in driving the Industrial Revolution.

It was these thoughts that spurred me into this journey to find the soul of the Devon, a personal odyssey if you like, to understand its beating heart. But while I adore my river, it is no more remarkable than any other Scottish river, and this river's tale is one that could be applied to any river in the land. The fact is all rivers are special, arteries of life that wind through our landscape, havens for nature and places for all of us to enjoy.

Chapter 1
RIVER OF SURPRISES
(June-September)

A soft dawn light crept over the brow of the Ochils that summer morning, revealing shadows and previously unseen gullies on the white-misted scarp. Despite the hour, the air was mild, and I sat with tense anticipation by the banks of the Devon, waiting and watching, hoping to catch a glimpse of beaver plying through the water. Sweet dreams are made of these – hope over expectation.

I had arrived in the rising light, stumbling in the poor visibility under cursed breath, but now the waiting was about to begin. It was a long shot, but even if no beaver materialised, then I was bound to see something else, for dawn is such a special time, when much of nature awakens but many other creatures in the natural world fall asleep.

The heady aroma of the river filled my lungs and I scanned the water repeatedly, my senses tuned to the slightest movement or change, a lone and hunched figure under the gathering sky. Sometimes the water swirled in an unusual way, the smallest quivering, and my interest was immediately aroused. But it always turned out to be the natural eddy of the current, or a moving trout below.

An azure-glinted kingfisher zipped upstream, and then later it flashed back down, a piercing cry forewarning its passage. A roe buck emerged on the far bank by the edge of an oat field. It was wary and picked up my scent and bounded away into a thick stand of nearby trees. But I saw no beavers, just the rippling rises of trout and a small party of long-tailed tits

Long-tailed Tits

bounding across the riverside alders.

I was undaunted, for that is often the way of things when watching wildlife. Besides, the softness of the air and the whispering flow of the water had calmed my soul like a soothing tonic, and for that, I was truly grateful. Anyhow, there would be other opportunities, and I vowed to return the following morning,

Alder Tree

and no doubt the next evening, too, and perhaps even the one after that, because the arrival of beavers on the river intrigues me like a burning passion.

There are many reasons for this fascination, but if I were to put my finger on it, then it is because they are a symbol of the past, our natural heritage, and of humankind's ability to destroy it, but also of nature's ability to bounce back if given half the chance to do so.

Their new-found appearance on the Devon is just one of many reasons why I adore this river. It is a place that forever has the capacity to surprise at every turn, whether it be the fleeting glimpse of an otter or the bow wave of a salmon surging over a shallow riffle.

For several years I had been aware that beavers were slowly colonising the river. A report of a road killed animal near Tillicoultry was an early herald, and with each passing year I stumbled upon increasing signs of small willows and alders felled by the bankside, leaving behind distinct, conically gnawed tree stumps. Indeed, I was becoming something of an expert at identifying their spoor, my eyes tuned to wherever even the smallest branch had been stripped of bark, the light glint of the underlying wood catching my eye.

Beavers have an inherent capacity to wander far and wide within river systems, although they are generally disinclined to wander long distances over land and cross watersheds to colonise a new river catchment. They are water animals at heart and feel an unease when any distance from a river, burn or loch. Despite this, those on the Devon most probably originate from the Tay and Earn river catchments where the animals have spread at an astonishing rate, following unauthorised releases in the early 2000s (or possibly even before then). They are now found throughout the Tay area and its many associated rivers and tributaries and number in their hundreds.

The most feasible route, in my view, for colonising the Devon, is from the Ruthven Water near Auchterarder and up through Glen Eagles and down into Glendevon. At the watershed, there is only a small distance across relatively level ground to reach the Devon system. Or perhaps the animals routed down from the head of Loch Earn into Loch Lubnaig by Strathyre in the Trossachs and then along the River Teith, or they could have come down via the Allan Water and into these more central parts of Scotland. I have also heard the belief that the Devon beavers were directly introduced into the river, which is an added possibility, although it is my inclination that they most likely arrived by their own accord, for their very survival depends upon a certain nomadic instinct that enables them to colonise new areas.

Whatever path they took, these special animals are back where they belong, following many centuries of absence after being hunted to extinction for their pelts and 'castoreum', an oil extracted from sacs near the base of their tail used for perfumery and their alleged power in curing ailments. Most beavers disappeared from Scotland by the twelfth or thirteenth centuries, although it is thought some still hung on in the Loch Ness area until the early sixteenth century.

In all my wanderings by the river, I had yet to observe one, for they are shy beasts, frequenting its quieter stretches and mainly nocturnal in habits. Besides, colonisation had only just begun, and I suspected that there were still only a handful of animals present along the whole river's course. Determined to see one, I began to frequent the more likely-looking middle stretches of the river at dawn and dusk from late June onwards, treading carefully along the bankside, lest I should inadvertently alert one of my approach.

A further dawn visit at a slightly different location revealed the first encouraging signs: willow branches stripped of bark and a small trunk protruding from the water that had been partially gnawed. I cautiously made my way further along the bank and heard the strange reeling call of a grasshopper warbler in amongst the thick luxuriant

Grasshopper Warbler

growth of meadowsweet and rosebay willowherb. It was unusually late in the season for a grasshopper warbler to still be singing, and normally, I would have investigated the source of the sound in the hope of catching a glimpse of this elusive and fickle bird, but this was not the time for such a diversion, for I was completely focused on my beaver quest.

Rosebay Willowherb

The grass here was long, there was no path and the going wet, with my trousers soon becoming so drenched from the morning dew they clung uncomfortably to my legs. On turning a bend in the river, I glimpsed a small willow that had been completely felled below the bank. Bark chippings from the gnawing action of a beaver lay by its base, and there was a grass-flattened track where the animal had made its way out from the water and up to the tree. I walked the bank further, but there were no more signs, so I retraced my steps back to the tumbled beaver tree.

Now was a real opportunity, because in my small rucksack was a sensor-operated trail camera I had packed in case of such an eventuality. This infra-red camera technology is incredibly useful when studying wildlife, for it reveals so much that would otherwise remain unseen, and I use trail cameras frequently when studying foxes at their dens and badgers at their setts. The concept is simple – the camera is carefully sited by an animal trail or other likely-looking spot and then left in place for several days with it being primed automatically to take a photograph or video when a sensor detects the changing heat signature of a passing creature. The equipment also means that photographs or video can be taken at night without the need for a flash or other artificial light, leaving the animals undisturbed.

I investigated several small willows around the gnawed beaver tree, looking for a suitable one to strap the camera to. None was ideal, but I eventually picked a curved branch that gave the trail camera a partial view down the beaver track and into the river. The bank was steep here and the pathway provided a convenient access for a beaver onto land, thus, there was every chance it would return and resume more felling over the coming days.

Later that week, the skies opened, and torrential thunderous rain fell for the best part of 24 hours, so unusual in its intensity for the time of year. Worried that my trail camera would become submerged and irretrievably damaged by the rising river, I returned the following dawn, only to find my worst fears realised with the camera completely under the angry, surging mud-brown water, which had been rising inexorably with each passing hour.

I could just see the top strap of the camera which was secured to the willow branch below. Retrieving it was going to prove challenging, given that the bank was steep, and one stumble could prove disastrous in the raging torrent. I gripped onto the branch of an adjacent willow and leant over as far as I dared, and with my other hand scrabbled under the water and managed to release the camera from its hold, filling my wellies in the process and receiving a sleeve-full of cold water. Surprisingly, when I examined the trail camera on the bankside, it was still working, the sealed cover having just managed to keep the rising water out. Relieved, I packed it away for examination later. As I did so, a fluttering in the damp grass by my feet caught my eye, followed by further movement nearby.

I looked closer and discovered two young sand martins stranded on the bank top by the raging water's edge, fully feathered, but still unable to fly. As the churn of rising water encroached upon their nesting burrow in the bank below, they had been forced to flee – a do or die situation for these little bundles of feathery fluff, half-fluttering, half-crawling up the steep bank to escape the maelstrom torrent.

Sand Martins

One of the birds looked in good shape, but the other was bedraggled, so I gently picked it up and transferred it to a safer position in the hope its parents would be able to find it and provide food. These youngsters were close to having the ability to fly, and if the parents were able to look after them for just another day or so, they would be safe. In the meantime, they needed to lie low in the hope that a mink or a crow did not discover them.

But they were the lucky ones, and as I looked at the surging water, it was apparent that many nesting burrows lower down the bankside had become submerged, their occupants succumbing to this watery hell.

It is always a game of chance for sand martins when nesting so close to the water's edge. But all was not lost, for other burrows on a high sandy cut were still well above the rising waters, little young heads peering nervously out.

I left the two juvenile sand martins on the bankside, and as I headed for home, a welling of emotion fell upon me, tears of concern about what the future held for them. So young, and yet in such great peril, not even having had the chance to experience the freedom of flight. These two little birds looked so fragile and vulnerable, and even if they survived this trauma, then many more challenges lay ahead, not least, a marathon migration to their wintering grounds in Africa.

For me, such international reflection is just one of the many magical attractions of the river; most inhabitants are resident, but some, like sand martins, sedge warblers and sandpipers are summer visitors, having arrived from sub-Saharan lands that are so different from Scotland – countries such as Guinea-Bissau, Mali and Cameroon, the Central African

Republic and Nigeria. These birds are true travellers, experiencing vast changes in the landscape on their travels, including the stark harshness of the Sahara Desert.

Later that day, I uploaded the contents of the trail camera onto my computer. In a space of 48 hours, four clips of video had been taken. I was thrilled to find that the first revealed a brief night-time glimpse of an otter swirling past the bankside, the second a wren flitting in amongst the grass, and the final two were the jackpots I was looking for: a beaver, laboriously making its way out of the water.

It was apparent that the camera had not been placed in the best position, as the beaver clips were close-ups of its back, but the images were good enough to discern the distinctive paddle tail of the animal. The clock on the camera revealed that the videos were taken just after midnight, confirming my suspicion that despite this stretch of the river being undisturbed, the beavers here preferred to be active under the cover of darkness. I replayed the video several times, excited by the capture. The most over-riding impression was the slow, cumbersome movement of this animal on land, compared with their much greater agility and turn of speed when in the water.

While delighted at the video success, it is no substitute for seeing an animal in the flesh, and with obsession now taking greater hold, I returned the following week when the flood waters had subsided. Again, no animals materialised, but there were other encouraging signs of activity: that partially gnawed willow I had previously found by the river was gone – a beaver had completed the task and taken the branch away.

Something else grabbed my attention, two beaver paths on the far bank leaving the water and into a thick perfumed-flushed patch of Himalayan balsam. Why had I not seen these trails before? My tracking abilities were perhaps not as adept as I had previously thought.

Several days later, I was back on the river once more and positioned myself by the bankside opposite those newly found trampled trails in the balsam. I watched the sun slowly fall through the sky until it slipped below the curvature of a hill. I love dusk and dawn in equal measure, soft and fading light, calming winds and gentle air, a time when nature envelops every sinew in your body. The smell of the river dominated my senses, humid air with a hint of earthiness.

Relaxed, my mind wandered. Then, there was a rippling disturbance in the water by the nearside bank about 100 yards downstream from me. I leant forward and more ruffled water appeared. Could this be a beaver at last? The ripples disappeared, leaving me to conclude they were made by a large trout working its way along the bank edge in search of food. Excitement and disappointment, they are such close bedfellows. That is the essence of the twilight, always wildlife activity about, and when one's senses are so finely tuned, even the most minimal movement is detected.

I stayed for a while longer, contemplating beavers and their place in our environment. Their return to Scotland has not been without controversy, with fishery managers worried about the impact of dams on migratory salmon and trout as they make their way to their spawning grounds upriver, and landowners concerned by felled trees, and exasperated farmers by beaver dams blocking their field drainage systems and flooding fields. They may also damage crops.

Scientific studies from other countries with healthy beaver populations suggest that while their dams may sometimes impede to a limited degree the passage of migratory fish, the overall impact is probably negligible, especially since dams are often ephemeral and fish can usually find ways over, through or around them. Furthermore, main river courses are unlikely to be dammed in Scotland because they will simply get washed away after prolonged rain, with beavers more likely to focus their damming activities on small slow-flowing burns on areas of ground with a gentle incline.

Beavers build dams to create expanses of water where they find it easier to forage, and to control water levels, so that the entrances to their burrows or lodges are well hidden by being submerged under water. Typically, beavers live in bankside burrows and they often create canals through thick riverside vegetation or soft mud in their home areas to aid their passage.

Beavers bring a plethora of environmental benefits: their dams and canals create pools and backwaters where invertebrate life can flourish, providing food and shelter for fish and other creatures, thus enhancing biodiversity. Their dams can also slow the flow of water, preventing flooding in areas downstream. They coppice and fell trees, letting sunlight filter to the ground below, enabling wildflowers and other plants to thrive. The environmental benefits they deliver are immense.

Beaver Dam

Indeed, during the same period when I was seeking out the Devon's beavers, I had stumbled upon a beaver dam near Dunkeld in Perthshire. In a wondrous feat of engineering, the beavers had dammed a small trickling burn, having placed gnawed branches carefully on top of each other, and then building it outwards, produced a formidable barrier. The intricate dam backed up the water, but a decent flow still spilled out from its lower part, maintaining the progress of the burn.

I hunkered down and peered into this dark, languid pool, a shimmering oasis lying deep within thick willow and alder carr. Pond skaters glided across the surface and a whirligig beetle swirled around in crazy fashion. By a shallow margin, caddisfly larvae lay on the bottom, encased in their protective homes constructed from tiny pieces of pond detritus.

Pond Skater

Whirligig Beetle

This pond was awash with nature, a haven for a multitude of invertebrates and water plants, all flourishing because of the industry of these beavers. Amphibians and water voles often colonise such pools, creating a whole new web of life. For me, this little Dunkeld beaver pond was compelling evidence of the ecological benefits these animals bring to our landscape.

Research by Stirling University found that beavers have an important impact on the variety of plant and animal life in their locality, with the number of species found in beaver-built ponds being 50 per cent higher than in other wetlands in the same region.

It is not just for ecological reasons we should welcome beavers, there are moral ones too. Beavers are an integral part of the Scottish environment and it is their right to be here, and they also have a right for us to protect them. Nonetheless there are two sides to every argument, and I understand the concern of some farmers, the people who put food on our plates.

Accordingly, while the current legislation in Scotland where beavers are protected, but with scope for management in areas where they are deemed to be causing a problem, appears to strike the right balance, there is already concern over how such control is interpreted and implemented.

While much of this includes common-sense mitigation measures, where funding and advice from Scottish Natural Heritage (SNH) can help alleviate issues in local areas without harming beavers, there is also the provision for lethal control if deemed necessary and as an option of last resort, and which can only be carried out under special licence. These situations include where there is serious damage (or the risk of it) to prime agricultural land, and where alternative mitigation measures either have not, or will not, address the problem.

Unfortunately, the most recent figures show this is not an option of last resort, with SNH having issued licences for 87 beavers – around a fifth of the Scottish population – to be shot in Tayside in the months following the Government's May 2019 decision to give beavers protected status in Scotland. Culling on such a large scale is a national disgrace, which shames us all.

As a conservationist, needless culling seems instinctively abhorrent to me, especially where beavers are concerned, since it would result in the killing of a whole family group. Furthermore, removing beavers, whether by culling or translocation, from a so-called 'problem area' is only a sticking plaster solution, as the vacant territory would no doubt be soon re-colonised again. Much better for concerned farmers and landowners to come to accept beavers on their land and recognise the environmental benefits they bring. As a non-landowner that is easy for me to say, but public opinion in support of beavers also deserves recognition. Beavers are part of everyone's natural heritage; they belong to all of us.

Garnering such a sea-change in opinion among some will take time, but for conservation to work, it is crucial that all elements of society – and their views and concerns – are brought on board. As such, in the short-term at least, management of one sort or another where beavers are deemed to be causing localised issues is unfortunately inevitable, though hopefully kept to a minimum.

I ventured back to this part of the river several more times in my quest to see beavers. On my last visit, sand martins wheeled over the river and a chiffchaff flitted in amongst the tangled branches of an overhanging willow. Perhaps among these sand martins were the two youngsters I had previously found, cold and scared after evacuating their nest burrow from the rising river, but now free and full of youthful exuberance. Darkness began to fall, and the first bats emerged. They swooped low over the river's surface, which made me think they were Daubenton's bats, which often fly in such a dare-devil manner.

As ever, a beaver failed to emerge, and with the rounded spurs of the Ochils now silhouetted against the fading pastel heavens, I suspected my time to see one must wait. No matter I thought, for these last few weeks of early mornings and late evenings sitting by the bankside had brought me closer to the very heart of the river and what made it tick. As I gazed up once more at the shadowy outline of the hills etched against the sky, my mind turned towards the river's source, which lay just over the near horizon. That was where I should explore next, the river's beginning, the little trickle that would soon turn into a mightier thing: the very gloaming-water flowing by my feet, and which is home to so much wildlife and so much wonder.

Chapter 2
HILL BURNS AND A JOURNEY
TO THE SOURCE
(June-September)

The hike from the foot of Alva Glen was proving enjoyable, following the hill track that skirted beneath the conical top of The Neb, then past Ben Ever and onwards into the heart of the Ochils. If one is looking for a quick and convenient access to the high ground of these fine little hills in central Scotland, then this is the route to take, for the incline is gentle and the going easy on the legs.

Tormentil   *Bird's-foot Trefoil.*

The summer sun shone warm and a gentle breeze rustled the rushes in the damp margins, while yellow-flowered tormentil and bird's-foot trefoil adorned the track verges, which in turn attracted many butterflies. Small tortoiseshell and peacock butterflies were the most frequent, but there were also several painted ladies, one of which alighted on the track, providing me with an excuse for a rest by squatting down onto my haunches to examine it more closely.

It may only have a wingspan of a couple of inches, yet these same paper-thin wings had carried this butterfly all the way from the desert fringes of North Africa to the spot where I was now standing in the Ochils.

Painted ladies are forever a source of wonderment: how can they fly so far on such fragile wings? I do not know the answer, for it is just one of nature's many mysteries. But it is certainly a challenging journey for these beautiful insects, and an individual I discovered the previous week displayed frayed and tattered wings, a testament to the punishment of the long flight.

Painted Lady

I left the butterfly in peace and continued up the track, soon arriving upon the southern fringes of Alva Moss, a vast boggy upland plateau, scarred with peat hags, and which stretches between Ben Buck and Blairdenon Hill. It was undeniably bleak at first glance: flat and featureless, but nonetheless holding an inner beauty that was hard to fathom.

I stood for a while, absorbing the scene, and trying to comprehend why this place was so compelling. Yes, it was empty, but experience of such places over the years has told me that this is only ever a veneer, because once you delve deep, nature's riches will soon sing out at you.

This moss, at an altitude of around 1800 ft, is effectively the dividing line between the burns that drain to the south and those that do so to the north. Unlike a true watershed that drains into different river catchments, these burns all flow into the Devon because the main river flows in a loop heading north as the diminutive beginnings of the Finglen Burn from the plateau, then east through the Upper Glendevon Reservoir and Lower Glendevon Reservoir before veering south and finally westwards along Strathdevon past Dollar, Tillicoultry, Alva and Menstrie in Clackmannanshire and out at the inner Firth of Forth at Cambus. Thus, some burns drain north into the early beginnings of the Devon, while others flow south to join the river when it is more mature and flowing parallel to the southern scarp of the Ochils.

Indeed, such is the peculiar looping course of the Devon, the source lies only five-and-a-half miles as the crow flies from its estuary at the Forth at Cambus, although the river itself is around 33 miles long, draining a catchment area of approximately 75 square miles.

The search began for the Devon's source. I knew from previous visits it was somewhere near, but the featureless landscape meant it took a couple of minutes to find my bearings as I trod in a zig-zag pattern over the area where it lay. Soon, I stumbled upon it, an oasis in the mind; in reality, a green and boggy sphagnum-filled pool. Just as how first impressions of Alva Moss can disappoint, then so too does the first emergence of the Devon, for it is a puddle, and no more than that.

Nevertheless, I'm sure the source of every river is underwhelming, whether it be Amazon or Nile, or Tay or Tweed, as the true beginning can only ever be the tiniest trickle of water, a mere seepage in the landscape, but one which like a snowball rolling down a hill, gathers pace, size and momentum. And, that was the magical bit, imagining what this insignificant pool was set to become, like lighting a blue touch-paper of unfolding greatness.

So, I let my mind drift, bringing images into the consciousness of the lower reaches of the river, the magnificence of the waterfalls and gorges at Rumbling Bridge and the Cauldron Linn further downstream, and the river's sweeping beauty as it meanders its way to the sea. Fuelled by such thoughts, this green sphagnum puddle suddenly became an exciting place; somewhere full of promise and the beginnings of a river and new life.

I gathered some sphagnum in my hand and clenched my fist tightly, watching the water ooze through my fingers, pure and clear. It was the start of the river's journey, but not the beginning of its story, for that goes back to the dawn of time, a period when the land was in tumultuous upheaval and violent eruptions shook the ground in unimaginable fury.

The name 'Ochil' is derived from the ancient Celtic word 'uchil' meaning the 'high place', and the very nature of these uplands originates from a dramatic series of geological events that stemmed from volcanic activity in the Devonian period some 360 to 410 million years ago – a time of great geological turmoil characterised by long spells of violent volcanic activity.

These lava flows poured out from underground vents, interbedding with resultant mudflow conglomerates and ash layers, building up over time, one layer covering the previous one, creating a great thickness of rock. It is a process I find difficult to comprehend: the sheer enormity of a land being sculpted by forces from within the inner earth – and one can only imagine the acrid smell of sulphurous fumes, the surging lava glowing at night and the terrifying noise of thunderous ground movements.

The dramatic steep southern scarp of the Ochils of today, which dominates the skyline of much of central Scotland, is the result of a fault that developed in this raised rock formation, with major earth movements separating the hard volcanic rocks to the north from the softer coal-bearing sedimentary rocks to the south.

In more recent times – and by that, I mean the last two million years – periodic ice ages helped sculpt the landscape further, with glacial ice sheets broadening and deepening existing valleys such as Glen Eagles and Glendevon in Perth and Kinross, and rounding off the tops of the surrounding hills. Once the ice began to recede for the last time about 10,000 years ago, gushing meltwater streams deposited sands and gravels on the lower ground. As the land rose due to the relieving of the tremendous downwards ice pressure, and sea levels dropped, the consequent alteration in the landscape resulted in the impressive Rumbling Bridge gorge. Previously, the upper Devon had flowed east to Loch Leven, but in a phenomenon known as 'river capture', the Devon changed course to flow westwards into the Forth.

Thus, it was the creation of these volcanic uplands, combined with subsequent glacial grinding forces and meltwater outflows, along with the continual onslaught of weathering, that lay the foundations of the course of the Devon.

I lingered by the source for a while longer and examined more closely the sphagnum in my hand – a sponge in effect, soaking up the rain and water run-off from this highest part of the plateau. And, like a sponge, water continued to slowly seep out, a lethargic dribble. I looked around and saw other dazzling green patches of sphagnum and dark, shallow peaty pools, which made me realise that there were several discernible sources of the river all around me, rather than just the one.

My fingers clawed at one of the peat hags, damp and thick and composed of a multitude of generations of slowly decomposing heather and other vegetation. Like the clump of sphagnum so recently in my hand, it is another sponge, which continually releases water, drip by drip, even in the driest of summers, keeping the water flow going and supporting so much life downstream.

There must be thousands of tonnes of water all locked up on Alva Moss, feeding these early river beginnings. This boggy plateau is the powerhouse of the river, the generator that drives its gathering momentum.

In the distance, Ben Lomond and several other fine Scottish mountains stood proud, including The Cobbler, Ben More, Stob Binnein and Ben Vorlich. Golden eagles haunt these not too distant hills, which means they can see the Ochils and some birds must occasionally venture across here, especially youngsters, which have a habit of wandering in winter. My mind started to race once more. Hopefully, one day, a pair will set up home here.

What a magnificent thought, and if not golden eagles, then perhaps sea eagles will be the heralds, which is entirely within the bounds of possibility, for on two occasions in recent years I have seen these huge birds soaring over these hills, which probably originated from a recent reintroduction scheme on the east coast of Scotland.

I focused once more on the river's source and started to follow these early beginnings for a few hundred yards down a wide peaty gully. Sometimes a small flow of water could be seen, but before it had a chance to gather pace, quickly disappeared into more moss, rushes and peaty pools. Then a noise drifted across the breeze, the sweet sound of burbling water. The main flow had suddenly become more pronounced as other side rivulets of water seeped in from the peat hags. I was witnessing the birth of a river.

A pool shone out at me, so I lay on the ground and gazed into its depths, turning stones over with my hands in search of life. Tiny water spiders

Caddisfly Larva

glided across the surface and on the stony bed of this tiny burn were several cased caddisfly larvae, their soft bodies cocooned by an outer coat made from particles of vegetation and minute gravel grains. True, they were nowhere near as abundant, or as diverse in range of species as lower down in the river, but nonetheless these were the pioneers of new river life.

A movement flashed below me: a tiny larva with a three-pronged tail – a mayfly nymph. Something bigger then caught my eye: an altogether larger insect and swimming in a rowing action with two long legs. It was a backswimmer or water boatman as it is often known. I also spotted a clump of duckweed growing by the water's edge, a glimmering of dazzled lime green. This little pool only a few hundred yards downstream from the river's source was already brimming with nature and establishing its own web of life.

I raised myself to a sitting position and breathed-in the landscape. The sun shone; the air was warm and momentarily filled with distant 'cru--uk' of a pair of ravens flying overhead. A grey wagtail in an undulating flight bounded upstream towards me before veering away when it sensed my presence. Wonderful, this early river was already a place of natural intrigue, so full of diversity and interest.

Brown Trout

Would trout venture so far up this burn? I believed this was entirely possible, so I started to search more pools downstream, but my efforts proved fruitless. However, I was sure they were there, because I had previously seen small trout in several of the other high upland tributary burns of the Devon, which, raised an interesting conundrum.

This led me the following week to ascend Dollar Glen towards the Glen of Sorrow, where the Burn of Sorrow tumbles down the floor of this steep sided valley and is itself fuelled by smaller feeder burns from an arc of hills surrounding the glen – King's Seat Hill, Skythorn Hill, Cairnmorris Hill, Tarmangie Hill and Whitewisp Hill. Burn of Sorrow and Glen of Sorrow are such evocative names. They are perhaps a reference to tales of captive lovers, political prisoners and clan chieftains banished to Castle Campbell, which overlooks the head of Dollar Glen, and which itself was previously known as Castle Gloom.

My ascent from the bottom of Dollar Glen was spectacular, with deep-sided gorges and tumbling waterfalls, which became especially impressive just past Castle Campbell. I did not pause to look on my way up, as I was on a mission, and with lightweight fly-rod in hand, I passed the impressive Sochie Falls and entered the lonely openness of the glen. The path soon became narrow and indistinct. This is not a place where many people venture, such is the roughness of the ground.

I found a dark, peaty pool in the narrow hill burn, crawled closer on my hands and knees to keep my profile low and with a flick of the wrist cast a tiny barbless black fly into the peaty water. Nothing. So, I casted again; a ripple on the surface, but perhaps that was my imagination. I pulled the rod back and sent the fly in once more, and the rod quivered. A small trout has taken the lure! I almost landed it, but the wriggling fish shook off the hook just as I was about to grab it in my hand. No matter, for although less than four inches in length, the fleeting glimpse of its burnished flanks was as impressive as any trout I had ever seen.

For me, the brown trout is one of our most attractive creatures, a fish with a remarkable range of colours – a feast of vibrant hues and nuances accompanied by a sprinkling of striking blood-red spots. When glimpsed from above when swimming under the water, the trout is a drab looking fish, but once held in the hand it is transformed into a sparking jewel.

No two fish are ever the same, and trout from different lochs and river systems can vary in appearance, sometimes quite dramatically. Even within a loch there can be more than one distinct breeding stock. The fascinating aspect of these hill trout in the Burn of Sorrow is that their population is completely isolated from trout in other parts of the Devon catchment because the gorges and waterfalls below in Dollar Glen are impassable to fish. So how did they get there?

Several years ago, I asked a freshwater fisheries biologist that very question. He told me that while there is no doubt that brown trout have been introduced to several high-altitude waters in Scotland – especially lochs – many populations in other hill areas, especially burns, are almost certainly naturally occurring.

He believed it most likely that they reached there towards the end of the last Ice Age when our catchments were very different, and migratory brown trout were able to access such high waters via ice lakes and other routes available to them at the time. Thus, these isolated upland populations may have been present for several thousand years. Furthermore, they spawn at a small size, and because of the length of their isolation are probably genetically unique from other stocks, giving them significant conservation importance.

It is equally possible that these trout in the high Burn of Sorrow were introduced, perhaps by inhabitants of the nearby Castle Campbell many centuries ago. Whatever the case, they are geographically isolated, and thus vulnerable, given the limited area of water where they are found and their susceptibility to succumb to environmental fluxes, including climate change. Indeed, those in the upland Burn of Sorrow occupy a narrow watercourse only a mile or so long, and it would be tragic if these glistening glimmers of uniqueness were to become lost for good.

Following my hooking of this diminutive upland trout, I was tempted to cast my fly into the Burn of Sorrow once more. But I had achieved my aim. Besides, these hill trout were so very special, and I was reluctant to disturb them anymore, so I dismantled my rod and headed back down the glen.

———————————————

Chapter 3
RIVER OF RESERVOIRS
(August)

Mid-summer, but the air was cool and grey clouds loomed above the calm waters of the Lower Glendevon reservoir in the heart of the Ochils, only a few miles from the source of the Devon that I had so recently visited. A couple of anglers in a nearby boat were enjoying a productive day and within the space of a few minutes, their rods twice curved downwards as rainbow trout eagerly snapped at their flies.

These anglers, however, were a fleeting distraction, for my eyes continually looked to the sky because I knew from experience that there was every likelihood a special visitor would turn up. Sure enough, not long after, a large bird appeared high above the dam at the far end of the reservoir – an osprey. On lazy, slender wings it soared higher and higher before swooping back down low over the water, flying this way and that, constantly scrutinising the surface in its quest for trout.

Osprey

It hovered, plunged downwards with a thumping splash, but then rose into the air fishless, before soaring over the ridge by Ben Shee to try its luck at the nearby Glensherup Reservoir. While the osprey had been unlucky this time around, one can only wonder at the intensity of the eyesight that is able to discern the shape of a camouflaged fish underwater and the skill and co-ordination then required to seize one.

Glimpsing the osprey was enthralling, and sometimes I still must pinch myself when I see one to ensure I'm not dreaming. When I was a child, ospreys were rare, but in more recent times they have made a quite remarkable comeback, and despite their new-found familiarity, I never tire of watching them.

With the osprey gone, the reservoir was quiet once more, save for the angling boat working its way methodically along the shore. It was a pleasant scene, loch against hill, the gentle peeping of meadow pipits and a pair of crows tumbling in the sky above me.

This reservoir is not all that is seems, for it is part of the Devon that has been dammed to provide water for the citizens of Fife. That little tumbling burn that developed from the source not so far from here, and where I had scrabbled around in search of caddis larvae and backswimmers, had been transformed into a huge body of water.

In all, the river and its tributaries in this upper part of the catchment fill five reservoirs, which in sequential order are Upper Glendevon Reservoir, Lower Glendevon Reservoir, Glensherup Reservoir and Glenquey Reservoir, and finally Castlehill Reservoir at the bottom end of Glendevon near the village of Muckhart. From Castlehill, the water spills over the final end-dam where the river resumes its course once more, bubbling over rocks and swirling around pools, the inconvenience of being interrupted seemingly a distant memory.

Glensherup was the first of the reservoirs constructed – in 1880 by Dunfermline Corporation waterworks as part of the water supply for the Dunfermline area – followed by Glenquey in 1911 to augment supplies. Lower Glendevon, known locally as Frandy Dam, was built between 1917 and 1924, with German prisoners of war helping in the initial construction stages, with the reservoir at the time being required to help supply the naval base at Rosyth. Upper Glendevon was built between 1950-5, by Fife County Council to augment the supply from Lower Glendevon. Backhill

Farm was submerged during the impounding of the reservoir in 1954 and the remains of the farm and farmhouse reappear whenever the reservoir level drops, an eerie reminder of people who once lived below these waters.

Sitting in quiet contemplation by Lower Glendevon Reservoir, I reflected on its place in our recent history, and of prisoners of war toiling on the dam to help provide water for the might of the Royal Navy. What stories remain untold behind the graft of this monumental dam construction? Countless, I am sure.

The final reservoir, Castlehill, is a concrete arch gravity dam and was built between 1972 and 1978. It is mainly used to provide compensation flows to the Devon. In other words, releasing controlled amounts of water to maintain the river flow downstream from the dam. There is also a water treatment plant at Castlehill.

However, despite the compensation dam, the flow immediately downstream from Castlehill Reservoir is more benign than it would naturally be, and even during periods of heavy rain, the river here seldom turns to spate. Does this matter? Well, yes it probably does to a certain degree, especially for salmon much further downstream, which like a good flow of water when migrating into the river. In saying that, the numerous tributaries of the Devon downstream from Castlehill Reservoir quickly fill the river, so that by the time the middle stretch of the Devon is reached, it once more behaves more like a typical spate river.

Despite the river being in its early stages, it had already been transformed beyond natural recognition by these reservoirs. Some might call it a form of ecological vandalism, but then again, everyone needs water, and to that end, this river, like so many rivers, is proving its importance to humankind. So, despite my initial reservations, I like the concept – rivers being so essential to our wellbeing, because it means we are reliant on rivers, and with that being the case, then we are so much more likely to look after them.

Such thoughts made me look out onto the water of Lower Glendevon Reservoir once more with renewed interest, watching the anglers cast their flies. The rainbow trout in the reservoir were not native, having been introduced from North America, so that was an additional element that was not natural. Nonetheless, these fish do give benefit through offering recreation and food for anglers, as well as providing thumping-sized prey for local ospreys.

Even the early youthful stages of the river above these reservoirs cannot be described as natural in the truest sense of the word. The ground all around is heavily grazed by sheep, which must impact upon how water seeps into the burn, both in terms of speed and water chemistry. I imagine that after the last Ice Age, this plateau would have been largely tree covered: birch, willow, aspen, rowan and the like. But since Neolithic times, the landscape would gradually have been cleared.

This made me wonder about the first people who lived by the upper reaches of the Devon and its tributaries, and so, a few weeks later, I ventured to the head of Borland Glen just below the top of Sim's Hill, which lies sandwiched between Glendevon and Dunning Glen slightly further downstream from Lower Glendevon Reservoir. The view down the narrow glen was magnificent with the tops of Innerdownie Hill and Commonedge Hill framed in the distance by a series of interlocking hill spurs.

It was not the view for which I had come, but rather the three large boulder-like stones lying mostly hidden in a patch of rushes – the remnants of a stone circle, a haunting place, made even more so by the sun setting over the brow of the hill. Not far from where I stood, a small tributary burn ran down a grassy gully, making its way to Glendevon and into the main river, and then on to Castlehill Reservoir.

A teardrop flecked my cheek – who were these people from prehistoric times who once stood on this very spot? What were they like and how did they live? It was a most humbling experience, made even more so by the nearby wind turbines above Glendevon – old meeting new, distant past and modern-day present all within the same panorama.

I stayed for a while longer, reluctant to leave, but it was getting dark, and it was time to go. As I made my way back down into Glendevon, my mind was alive with wild imagery of the landscape as it might have been then.

We can only speculate how this part of the Devon catchment looked at that time, although it probably consisted of a patchwork of forest and clearings. There is certainly archaeological evidence of farming in Scotland from around 3,700 BC, so it seems likely that land would have been cleared to some extent, both for food production and wood for the building of shelters and fuel. The primitive field systems of this era would have been concentrated on areas of better drained ground in the hills.

One would also imagine that this stone circle at Borland Glen would have been open to the elements, and not forested, given that that such sites were most probably used for religious, ceremonial or social purposes, and influenced hugely by the heavenly open views of sun, stars and moon.

The Devon catchment holds other prehistoric relics from Neolithic times and onwards, some still obvious to the eye, but most obscured by the course of time. Indeed, it is thought that the intensity of farming that peaked in the eighteenth century has obliterated many of these ancient archaeological signs.

The more obvious nearby prehistoric landmarks include the Gray Stone south of Dunning, a prominent standing stone and another one near Dollar, below Hillfoot Hill, called the Castleton Stone. Intrigued, I ventured to the Castleton Stone one afternoon, and on close inspection, 'cup' marks could be discerned engraved on the surface – a typical feature, but the significance of which is unknown.

From these early times, the upper Devon and its high tributaries would have been a vital source of water for people. But what about it being a source of food in the form of brown trout? I find that unlikely, for the trout in these upper reaches of the hills are generally small, and the impassable rocky gorges further downstream mean that no salmon have ever ventured up into these parts. So, trout would most likely have been a culinary treat rather than a staple, caught by spears or in simple traps. Further downstream, where brown and sea trout abound, and salmon run the river, fish would undoubtedly have been a treasured resource.

***

Chapter 4
THE RIVER'S ENGINE ROOM
(August-September)

Water rushing against wader-clad thighs, background hills and sweeping pasture. I was wading in the upper river in early autumn – the stretch between Castlehill Reservoir and the Crook of Devon – sweep-net in hand and furiously kicking at the rocks and small stones on the river's bed, catching the detritus stirred by my flailing boots. Accompanied by Alan Graham of the Devon Angling Association, we were 'kick sampling' to determine the abundance and variety of invertebrates, which provides a valuable indicator of the health of the river (part of the national Anglers' Riverfly Monitoring Initiative to provide data and to ensure that any pollution events that affect biodiversity are quickly detected).

Ever since I had moved to the Devon area, I had enjoyed fly-fishing along its course, but now it was my turn to give something back, to help learn more about the river, to monitor it and safeguard its precious inhabitants from potential threats.

We emptied our haul into an examination tray. At first, we could only see a handful of bugs in amongst the mix of gravel, small stones and weed fragments, but then suddenly the tray became transformed into a wriggling mass of life as a whole host of tiny creatures emerged.

We looked down in a sense of wonderment, for the riverbed that at first glance appeared to be just bare stone and gravel, was packed full of an incredible array of invertebrates. In the tray were an abundance of mayfly larvae (nymphs), and those of caddisflies and stoneflies, as well as tiny freshwater shrimps and a cornucopia of other life including beetles and worms.

These lifeforms are the engine room of the river, the driving force that supports birds, fish and so much else. Trout and minnows depend upon this natural bountiful larder for their very being, both when these invertebrates are in their larval stage, and when the mayflies, caddisflies and stoneflies miraculously emerge as fully winged adults.

For the mayfly, it is a case of beauty and the beast, as the underwater larval nymphs look fierce and ugly, but once the adults emerge on gossamer wings, they have incredible grace and elegance. These invertebrates, however, are much more than just food for other creatures. The benefits they provide are diverse including decomposition and the recycling of nutrients, so important for a healthy and productive environment. In short, without these mayflies, caddisflies and stoneflies, as well as all the other invertebrates, the river would be a wilderness, a lifeless flow of water draining the land, but doing little else.

I plucked a stonefly nymph out from the tray, placed it on my finger and examined it more closely. It was a large one, almost half-an-inch long, with a long two-pronged tail, segmented abdomen and flattened upper body. The compressed body is a key adaptation of many nymphs, as it provides streamlining against intense water flow so that they do not get swept away and enables them to shelter under stones. Furthermore, smaller nymphs avoid being swept away by living in the thin veneer of low velocity water, known as the boundary layer, that surrounds stone surfaces.

The body of this stonefly was armour-plated in appearance, its body protected by a hard exoskeleton, which would prove useful protection from other underwater invertebrates, although I doubt it would be enough to deter a trout.

Stonefly Nymph

This stonefly was a good find, for they are especially sensitive to water pollution, so where they occur, all is well with the river. Like the mayflies and caddisflies, they spend virtually all their life – up to two years – living on the riverbed, grazing algae and vegetarian detritus, or preying upon other tiny creatures, depending on species. Then, they crawl out of the water and moult one last time, unveiling from within this nymphal skin a winged adult fly. It is a true miracle and I find the physiological changes that take place, transforming from one form of creature to another, a spellbinding wonder of nature.

Once they become full-blown adult flies, they are on the wing for only a very short time (perhaps a couple of weeks), their sole purpose to mate and lay eggs back in the water for the life cycle to begin once more. The adults then die, their lifeless floating bodies providing a bonanza for trout to feast upon.

A few months before this river sampling expedition, I had ventured down to the middle river at Vicar's Bridge one evening to witness a mayfly hatch. I leant over the bridge and watched the drama unfold as

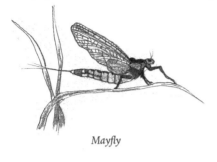

Mayfly

dusk took hold. At first there were only a few mayflies about, but as twilight enveloped the balmy air, hordes suddenly appeared, dipping up and down in an aerial dance. At the behest of some unknown cue, possibly temperature or the diminishing light, or perhaps a combination of the two, these flying wonders had emerged in a massive hatch. This aerial ballet was hypnotic to observe, almost as if these mayflies were suspended on puppeteer strings as they bobbed up and down in their rhythmic courtship.

Sometimes, I could see a male grab a female in flight and then couple together, ever so brief but naturally intimate. This was the moment, the whole reason for that two years living under stones on the riverbed, facing the dangers of trout or being eaten by dippers or other larger insect larvae. The adult females will then drop their fertilised eggs in the water, which fall to the bottom where they stick to plants and stones. The job is complete, and the adults perish – such a complex lifestyle for such a short and uncomplicated final act.

As I watched the mayflies, bats began to emerge on quivering wings to make good this aerial bounty. Sand martins and swallows are also heavily dependent on these regular emergences, swooping excitedly whenever they occur.

It is all part of the web of life – algae and plant matter to nymph or larvae, then onto dipper, minnow and trout, finishing with top predators such as kingfisher, heron and otter. So incredibly simple, but so fragile too, and always liable to collapse if one building block is removed due to external environmental influences, such as pollution.

Our invertebrate sampling over the period confirmed that the Devon is on the whole a healthy river, but like any other river, threats are ever-present – overspills from sewage treatment plants on the middle and lower stretches during times of heavy rain, agricultural run-off of fertiliser and pesticides, and unexpected incidents, such as happened on the Devon several years ago when an acid spill decimated the trout population over a stretch of the upper river.

There are all kinds of legal protection and enforcement for our waterways – including the Water Framework Directive – but unacceptable incidents keep on happening, which is why such voluntary sampling work to detect water changes by volunteers such as the Devon anglers is so important. For this reason, anglers are the unsung heroes and guardians of our rivers.

The kick-sampling of the river had opened my eyes to a whole new world; as exciting as any dramas played out on the African plains, a place where nymphs of various species engage in battles, preying upon each other, and themselves being feasted upon by trout, dippers and other creatures.

On a further sampling visit, I examined in more detail some of the other creatures caught. As well as the larval insects in the kick-sampling catch, two other invertebrates caught my eye – one of which was a leech that latches onto fish, and the other, a strange-looking freshwater crustacean, popularly known as a freshwater shrimp, but in scientific terms described as an amphipod. They, too, are key indicators of a healthy river.

Of all the creatures caught, caddisfly larvae were among the most fascinating, with there being two distinctive types in the tray – those with an external case made from tiny stones, sand grains, and plant material, and those without a protective covering; little wriggling six-legged bundles, with two short posterior hooks visible on the tip of the abdomen.

These cased caddisflies are real natural marvels with their protective covering of river detritus, stone and sand particles, glued together by a silk lining spun from special glands around the mouth. A bit like a hermit crab in its requisitioned shell, they move across the riverbed encased in their protective housings.

I turned over a further stone in the river shallows and discovered an unusual form in the life stage of the caddisfly, this time protected by a dome made from tiny gravel. These intricate domed casings covered the submerged stones much in the same way as barnacles adhere to rocks on the seashore. The fixed nature of these protective shelters meant that this caddisfly stage was sedentary, and I identified them as pupation shelters, nearing the final stages of their life journey prior to emerging from the water as an adult fly.

Despite being common, adult caddisflies tend to keep themselves to themselves and are not often seen. But one evening as an experiment, I placed a light-trap designed for live-catching moths by a small tributary burn of the Devon, and the next day it was full of adult caddisflies, underlining that they prefer to be on the wing at night.

This incredible productivity of the river was further brought home to me in early September of my river year when the Forth Rivers Trust invited me to participate in an electro-fishing exercise on the Dollar Burn – one of the tributaries of the Devon, which feeds into the middle section of the river.

Electro-fishing is an effective way of monitoring fish populations, especially juveniles, and is a specialist technique that temporarily stuns the fish, enabling them to be scooped up into a net before being later released unharmed back into the river. The Alva, Dollar, Menstrie and Tillicoultry Burns, along with some other tributaries of the Devon, are regularly monitored by the Trust to assess how fish populations are faring and to detect any trends.

Our small sampling team was headed up by Trust fisheries biologists Jo Girvan and Jack Wootton. It was a day of squally rain showers, and the water level in the burn was relatively high, but Jo decided conditions were still suitable and began electro-fishing. She waded up the burn sweeping a long-poled anode (which looked a bit like a mine detector) into the swirling pools ahead of her. Jack was right behind her and he quickly lifted into his net any fish affected by the electrical current, which were then placed into a pail of water. Fish that are missed, recover within seconds and dart back to the bottom of the burn.

A flash of silver in the churning pool. A cry went out: *"There's one, there's one!"* Up scooped Jack's net and within its mesh lay a glistening salmon parr. The young salmon was removed from the net and we briefly examined it before placing it into the bucket of water. It was a most attractive fish, no more than a couple of inches long with burnished flanks and a long slender body. It was hard to imagine that one day this could well turn into a most impressive salmon weighing several pounds.

We continued to electro-fish and many more young fish were caught. For me, it almost defied belief that this tumbling, shallow rock-strewn stretch of water, which to all intents and purposes looked devoid of life, could hold so many fish. They just kept on popping up every time an electrical pulse was sent through the water. If one peered into this burn from above, then there would be no sign of young trout or salmon, for they blend in so seamlessly with the bottom, the rippling water above helping to aid their concealment.

After a carefully timed sampling period, we had covered a 40 ft long stretch of the burn and returned to the bankside to examine the 'live-catch' held within the pail. Jo was satisfied with the result, with 39 young trout and six young salmon caught. The good number of one-year-old trout was especially encouraging, as it indicated that the most recent spawning season had been successful.

The trout and salmon were carefully measured and recorded before being released back into the water. How could such a small stretch of burn hold so many fish? Indeed, it was likely many more trout and salmon in that small section of the burn had been missed. The answer lay in this abundance of invertebrate life lurking beneath the water, a massive crawling larder of protein that sustains fish and so many other creatures.

Eel

We also caught a small eel, which caused a bit of excitement, as their numbers have plummeted alarmingly in Europe over the last few decades, with the International Union for Conservation of Nature (IUCN) listing the species as critically endangered.

How eels reproduce has been the subject of great debate since the earliest of times. Aristotle thought eels generated spontaneously from mud or slime and Pliny the Elder suggested that eels rub their skin against rocks and the pieces that came off turned into young eels. Isaak Walton in the classic work *The Compleat Angler* (1653) postulated that eels generated by "... *the sun's heat or out of the putrefaction of the earth*".

The reason for such confusion lies in the fact that nobody had ever discovered a mature eel laden with eggs or milt and it is only in the last 100 years or so that it has been deduced that eels in fact embark upon an epic 3,000 mile migration from our rivers to the Sargasso Sea off America to spawn. In what must rank as one of the most incredible of biological feats, their tiny glass-like larvae then gradually drift back to Europe on the Gulf Stream before ascending our rivers as elvers.

All kinds of reasons have been put forward for this decline in numbers, which some estimate at more than 90 per cent, including climate change, over-fishing and barriers on rivers impeding migration.

Eels are very tough creatures and can even leave the water at times to make their way from one body of water to the next, slithering across damp grass on dark moonless nights. Therefore, eels can sometimes be found in landlocked ponds and ditches where they survive in poorly oxygenated waters that other fish would find difficult to live in.

As we examined this eel in our water filled bucket, its body wriggling like a snake, I tried to envisage what the future held for it. This little piscine serpent had already been through so much, starting life in vast oceanic depths so distant from this rainy bankside in central Scotland, and drifting in the currents at the mercy of all kinds of predatory fish. Did it then reach the river's east coast estuary by travelling around the north coast of Scotland, or had it come up through the English Channel? And was this fish descended from other eels that once lived in the Devon, or did it just happen upon the river by chance? This eel was a true enigma, and with a sense of awe, we tipped the bucket and watched this wonder of nature slip back into the coppery waters of the burn.

Chapter 5
AN UNDERWATER EXPERIENCE
(September)

It is a truism that fish, mammals, birds and even insects gain all the plaudits and recognition when it comes to appreciating river life, with water plants tending to be overlooked. Water plants, however, have jaw-dropping beauty, and no more so than the superb run of glorious white-flowering water crowfoot I discovered on a long stretch of the upper river below Castlehill Reservoir, which sweeps all the way down to the Crook of Devon.

Water Crowfoot

I had always had it in my mind that water crowfoot preferred slower, more benign stretches of water, but here in the tumbling upper river, it was flourishing. Water crowfoot grows in long trailing clumps, which taper down the river current, waggling in the flow, and when I stumbled upon this proliferation, it reminded me of an English chalk stream, such was its verdant growth combined with the clarity of the water.

It was early September and many of the plants were still in full flower, their white petals centred by a yellow orb. I hunkered down onto my haunches to examine one of the plants in a shallow margin more closely. A member of the buttercup family, these flowers were so exquisite that they danced like little white gems upon the water's surface.

Here, it was growing like a mat in only a few inches of water, rounded leaves coating the water surface in a sheen of green. Further out in the middle flow, I could see more plants under the water, but such was the depth and the speed of the current, there were no flowers showing on the surface.

Intrigued, I returned the following week and donned a wetsuit, snorkel and mask, and then took to the river, a burbling of peaty water sweeping past my face as I plunged in. Suddenly, I was in a different world, and a surprisingly noisy place, too, from the rushing current that bundled and churned its way over rocks and boulders. The water was tinged brown from the peaty run-off from the hills, but the visibility was nonetheless excellent.

This is an unforgiving environment, forever under the high-pressure flow of the current; the equivalent of a land environment being continually pounded by gales. As I snorkelled, it was all too apparent why invertebrate nymphs have such flattened bodies, to avoid the worst excesses of the surging water.

My body caught a riffle, sweeping me downstream at what seemed like the speed of a torpedo. I grabbed a boulder to stop myself, swinging round so that my head was now facing upstream. That was better, I could relax now, which provided the opportunity to study these underwater crowfoots in more detail.

Incredibly many of their flowers were a few inches under the water. Perhaps this submerged flowering is done in the hope that a dry spell might lower the level sufficiently for the blooms to emerge into the breezy air above. Such conditions might only happen for a short spell, so better to be ready for the moment rather than lose the opportunity.

I also noticed that apart from the rounded, floating leaves at the top end of the plant, the rest of the leaves were comprised of thin tassels; an adaptation to cope with the fast water flow by reducing resistance as much as possible so that they are not plucked from the riverbed by the strength of the current.

The water momentum was strong, pushing me backwards all the time, a reminder of the sheer power of the river. I grabbed another rock slightly ahead and pulled myself forward against the water's flow. A trout darted away, there one second and gone the next. This was like being in an underwater forest, with these long trailing frilly tassels of water crowfoot providing a superb place for trout and minnows to rest and hide. I parted some of the waving weedy strings with my hand where in among the fronds tiny water snails clung tenaciously.

These tough water plants are a vital part of this stretch of river, providing shelter and homes for fish and invertebrates. Their importance here is further exacerbated by the lack of bankside trees along this part of the river, which is surrounded by heavily grazed sheep pasture. Overhanging willows and alders are ecologically important by providing trout with shade and insects to feed upon, and their protruding underwater roots offer nooks and crannies for all sorts of creatures to live. Thus, these water crowfoots were providing crucial shelter that was otherwise missing in this part of the Devon to a whole plethora of river life.

There were other underwater plants here too, including the long trailing green strands of alternate-flowered watermilfoil, which looks very similar to water crowfoot, and many of the rocks and stones were covered in submerged moss, providing even more diversity of habitat for the river's creatures.

I pushed away with my hand and was back in the middle of the river again where I gripped onto a rock to steady myself. Here, a small stickleback flickered ahead of me, using its side pectoral fins to scull through the water. It, too, showed superb adaptation, the three spines on its back acting as a deterrent from being gobbled by trout and water birds.

The flow here was strong, but this diminutive fish appeared to cope with the pull of the current with tremendous ease. However, its energy expenditure must be considerable, forever swimming to keep on station. But like the flattened insect nymphs, the vertically compressed body of the stickleback is well-shaped to minimise water resistance.

I released my grip from the rock, tumbled around and shot further downstream before encountering a shallow calmness, from where I floundered to the bankside where I sat for a while in the shallows, watching the undulating water catch the autumn sun so that it sparkled and danced like the reflection from a thousand mirrors. I felt exhilarated from the cold freshness of the water and the new aspect to the river that snorkelling had revealed.

Dripping water, I wandered back along the bankside, scanning the river margin for other plants. Despite the lateness of the season, the dazzling sky-blue flowers of water forget-me-not brightened one section of the bank. The name is believed to derive from German folklore where an armour-clad knight had picked a clump of the flowers for his lady as they strolled along the river. Unfortunately, the hapless knight fell in the water and just before he drowned, threw the flowers to his love, crying 'forget-me-not'. It is a wonderful tale for a most beautiful plant – a wildflower which the poet Samuel Taylor Coleridge eloquently described in his poem 'The Keepsake' (1800) as "that blue and bright-eyed floweret of the brook".

Monkeyflower

Monkeyflower caught my eye on this upper river bankside, too, an introduced species from North America with brilliant yellow flowers. There was also water mint, with a delicate dusting of lilac atop their little flower spikes. I plucked a leaf and popped it into my mouth, the minty flavour as strong as any shop-bought plant. By a calm watery margin, strings of algae coated the stones, turning into globular little balls in some places. Algae are the simplest of plants, yet so important in the river's chain of life by providing food for invertebrates to graze upon; the initial fuel for the invertebrate engine that drives the rest of the river.

A dipper zoomed upstream on dumpy wings and alighted on a rock in the middle of the river's flow, its white breast contrasting starkly against the darkness of the river. The Devon has a good population of these birds, a sign of the health of the river, for they are like natural barometers, their presence an indicator that there is plenty of underwater invertebrate life for them to feast upon.

Water Mint

Dipper

I watched the dipper as it plopped into the water and disappeared. The dipper is one of our most remarkable songbirds because of its ability to dive and forage under water for caddisfly larvae and other invertebrates. It is claimed that the dipper can walk under water, but this is not really the case. Instead, it uses its wings to exert downward thrust to counteract its natural buoyancy, enabling it to forage for small creatures along the bottom. So rather than walking, it is more of a swimming action using its wings as paddles with the feet clawing along the rocks and pebbles on the bottom.

They are superbly well adapted to the ways of the river. When the Devon is in flood it is a truly ferocious place where it would be impossible for dippers to feed. During such times, dippers move up the numerous tributary burns of the Ochils where the conditions are more benign and the underwater foraging is much easier, albeit less productive than the main river.

Their aquatic niche means dippers enjoy plentiful food all year. Unlike a blackbird or song thrush for whom worms and other creatures are hard to find when the ground is frosted hard, for the dipper, the movement of the river ensures it is always ice-free and there is food to be had, no matter the time of year. I have, however, noticed during extreme winter-chill that dippers swim on the surface out in the centre of the river more than usual, before diving under. I can only speculate that this is because their invertebrate food moves into deeper water at such times.

I revisited this upper stretch of the river a few weeks later, the water crowfoot no longer in bloom as autumn's chill began to take hold. Downstream from me, a pair of female goosanders bobbed about in the water. They had not spotted me, so I lowered my profile and brought them into focus through my binoculars. A smile broadened across my face, for as the goosanders swam, they continually dipped their faces under the water in search of trout – an equivalent to snorkelling, just as I had been doing in this same river stretch a few weeks previously.

Goosanders tend to get a bad press, with some anglers and river managers concerned they may impact upon trout populations. My years of visiting the Devon have convinced me that nothing could be further from the truth, for goosanders are an essential component of a healthy river. Besides, there are numerous other fish predators on the river – otters, herons, kingfishers and the like, and one more thrown into the mix is not going to make much difference. Even the humble dipper is a fish predator, eating small fry and no doubt devouring many trout and salmon eggs too.

Indeed, as my electro-fishing experience in the Dollar Burn has previously revealed, there are plenty of fish in the river to go around. After hatching, numbers of trout and salmon can be very high but reduce dramatically as they compete for the limited resources of space and food. Thus, predation during these early stages is simply removing fish that would not have reached adulthood anyway. Furthermore, the remaining fish have a higher chance of survival as there is reduced competition. That's the natural way of things and how rivers have functioned since the dawn of time.

If there is a desire by river managers to have as many fish in the river as possible, then a bottom-up approach to management is required, ensuring there is a clean habitat, with plenty of trees with trailing branches and complex root systems that protrude into the water from the river bank, offering places for fish and invertebrates to thrive.

It is so very simple, provide the right conditions and then all river life prospers. It is all about letting nature take the lead and looking after itself.

My snorkelling on the upper river had whetted my appetite so much for the riverine underwater world that I was keen to compare the contrasts with the lower river, so the week after I had first taken the plunge by the Crook of Devon, I found myself slipping into altogether slower flowing and calmer waters further downstream.

Whereas the upper river was a burbling frenzy of gin-clear tumbling churn, here on the more canal-like lower river, the water was murky and benign. I was immediately struck by the poor visibility, as well as the soft silty bottom in which I could plunge my fist in right up to my forearm without meeting any resistance. This silt was rich in organic matter and on several occasions strings of bubbles rose from little holes that miraculously appeared on the riverbed, which was escaping gas from the vegetative decomposition process.

I glided into a huge pool; the bottom no longer visible such was the depth of the silt-laden water. Then I suddenly hit a shallow bank, no more than a foot deep, and upon which a huge shoal of minnows flashed and swerved, their silvery dark-striped flanks catching the sun.

When one thinks of rivers, especially Scottish ones, trout and salmon are the fish that come to mind, but here on this stretch of the Devon, the minnow was king, and their numbers were quite astonishing. These fish clearly thrived in this slower-moving water environment, the riverbed rich in detritus and the alder roots

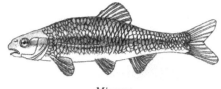

Minnow

and submerged fallen tree trunks providing productive places to feed.

The importance of these fallen trees, roots and branches immediately became apparent to me as I came upon a half-submerged willow that swarmed with minnows, flickering their way through this twisted entanglement of twigs and waterlogged wood. Just as how a rocky oceanic reef is an oasis of life, then so too are these tumbled trees in rivers, providing shelter for fish and invertebrates. They also help to slow the water flow in times of flood, providing additional protection for river creatures.

For kingfishers and goosanders, these minnows provide vital food. This raised a question in my mind. The complexities of how creatures interact is never straightforward, and I wondered whether goosanders benefit trout populations, by catching minnows that would otherwise compete with trout. Is the maligned villain of the river helping trout prosper? Now, that was a wonderful thought and one I was keen to throw into the mix when next discussing river ecology with my angling friends.

I floated over to the far side of the pool, the edge brimming with Himalayan balsam and tangled willows. I raised my face from the water and by the edge of the balsam, the familiar cone-shaped stump of a beaver-gnawed willow stump shone out at me, flattened vegetation all around. Beavers had swum in this very same pool and by lying half-submerged in the water, I was experiencing a beaver-eyed view of the world. I could also appreciate how beavers regarded water as a safe sanctuary, a protective place to be, for with just a flick of their paddle tails, they can disappear into the muddy depths of the river.

I snorkelled for a while longer, but despite the protective wetsuit, the cold was getting the better of me, so with a kick of my flippers I glided back to the bank where I heaved myself out of the water, the thick river silt trapping my flippers momentarily, before I managed to pull them free.

Chapter 6
ALIEN INVADERS
(October)

It might have been early October, but this bankside on the lower part of the river near Menstrie was still aflush with the sweetly-sick perfumed aroma of Himalayan balsam – an alien invader that is gaining an increasing stranglehold on the river with each passing year.

Although beginning to die back in anticipation of the onset of winter, many of these thick-stemmed plants were taller than me and their growth was prolific. I started to crunch my way through a thick stand, but quickly gave up, beaten back by the sheer density of growth. I examined one of the plants and there was no denying that their gaudy pink-purple flowers were rather attractive – and complicated, too, with the lower part of each deep flowerhead shaped a bit like a policeman's helmet. The flowers also showed great variation in colour, ranging from deep purple to lilac and almost white.

Himalayan Balsam

Bees and other pollinators find their flowers irresistible, and because of their depth, buzzing insects disappear completely inside, before emerging with their bodies coated in pale-powdered pollen. Indeed, when I first came upon bees feeding on Himalayan balsam many years ago, I thought I had discovered an unusual type of insect such was their ghostly coloration.

Himalayan balsam was introduced into gardens in the mid-1800s and soon escaped to colonise the countryside. It is particularly abundant along river courses because their exploding seed capsules (each plant can carry 800 or so seeds) can fling seeds a reasonable distance, which are then carried downstream in the water flow to colonise new areas. Unfortunately, their tall invasive growth shades out native plants and the die-back of extensive stands over winter can leave riverbanks bare and exposed to erosion, which can lead to siltation of trout and salmon spawning grounds. In addition, while pollinators adore balsam, it creates a further problem in that their brassy blooms detrimentally lure nectar-seeking insects away from native flowers, upsetting the natural equilibrium.

As I had discovered with the rainbow trout in the reservoirs in the upper reaches, as well as with the more benign yellow-petalled monkeyflowers further downstream, the river and its environs are awash with non-natives. Some are benign – such as monkeyflower – and thankfully, the environmental conditions are not suitable for rainbow trout to breed in Scotland, which means there is no danger of them spreading uncontrollably. Unfortunately, many other of Scotland's non-natives are bad news indeed.

American Mink

One of the most notorious is the American mink, and only a few months previously I had experienced a most unusual encounter with one. I was making my way along one of the banksides towards Vicar's Bridge on the middle river, while my Welsh Springer, Lottie, trotted ahead, sniffing every nook and cranny as spaniels are so wired to do. Suddenly, Lottie stirred a commotion by the waterside, and a dark furred animal, about the size of a ferret, scooted up a bankside alder and perched on a 15 ft high branch where it peered down at us nervously. It was a mink; I never realised they were such good climbers.

I was surprised this mink had not taken to the water instead, for they are extremely efficient swimmers, but perhaps Lottie had inadvertently cut off that escape route. The mink continued to stare down at us, claws gripping the bark so tenaciously that it obviously had no intention of descending until certain we were long gone. As I examined the creature through my camera lens, its sheer beauty simply shone out at me, with its cute weaselly face, little white chin and thick-furred body. The flipside is that the mink is not native to our shores – they hail from North America – and are descended from fur farm escapees from several decades ago.

Mink are universally detested by anglers and conservationists because of the havoc they wreak upon native wildlife. The mink is a supreme predator – an accomplished swimmer and trout catcher, it will also take waterfowl and waders, along with their chicks and eggs. They can devastate vulnerable sand martin colonies by being able to dig out their burrows. And, as I had just discovered, they are also adept climbers and probably prey upon birds' nests in trees, too.

Mink are also pivotal in the demise of our water vole populations, which have plummeted by more than 90 per cent in recent times. The mink is a natural carnivore and it is easy to demonise it, which is wrong, for the animal is no fiercer or more of a villain than any native fox, otter or stoat. It is only doing what evolution has made it do. It is not the mink's fault it is here; it is entirely ours, and just one example of many that illustrates our inherent capacity to interfere with nature to the detriment of the environment. The problem now is that the pace of human-aided colonisation of non-natives has gathered real momentum over the last few decades – and experience tells us that even the slightest tinkering can have the most undesirable and long-lasting environmental impacts.

At any rate mink do not seem to be as abundant on the Devon as they were ten years ago, possibly because otters, which are larger, are doing well and out-competing them. If this is the case, then it does show that some of our natives are capable of biting back.

Controlling these undesirable natives is always a daunting task. Volunteers from the Devon Angling Association assiduously strim down large patches of Himalayan balsam on the banksides to prevent it seeding and spreading further, and while such action slows the spread, it is very much an uphill battle, given its sheer ability to spread and colonise. The shallow roots can also be pulled out as a means of control, but that is very labour intensive. Herbicides are an alternative option, but not a desirable one in my view, as they can kill off adjacent plants and add pollutants to the environment.

Mink control is equally problematic. There is little point in having a mink eradication programme solely on the Devon because mink from other areas of central Scotland would simply move back in. Any eradication policy would have to be carried out on a much wider and integrated scale, which costs money and a lot of organisation.

But whether it be mink, Himalayan balsam, Japanese knotweed, or any other invader currently on the river, the potential impact of other new arrivals is equally frightening. The North American signal crayfish is an ever-present threat, which occur in the nearby River Clyde catchment, and if they were to appear in the Devon, they could pose a serious threat to trout and salmon eggs.

Similarly, increasing numbers of pink salmon – a native of the Pacific – have been turning up in northern Scottish rivers in recent years, after having been introduced into rivers in Russia in the 1960s. Our own native Atlantic salmon has problems enough, without the additional threat of a new competitor.

As I stood on this river bankside near Menstrie, engulfed by the towering balsam, I could not help but feel fearful for the future of the river with all these current and potential invasive threats; a ticking time-bomb of our own making, and one with the potential to cause great turmoil.

———————————

Chapter 7
TAKING TO THE WATER
(October)

I was in a quandary, for a large stretch of the lower Devon is relatively inaccessible by foot, surrounded by farmland where there are few paths, with the river banksides being steep and lush with thick vegetation. Over the summer and into early autumn, I had made several forays down to this part of the river, being able to access various limited parts of the bankside where tantalising glimpses of the slow-flowing water loomed out at me. On two occasions, flighty mallards took to the air on clattering wings, and another time at dusk, the soft light was brought to life by bats swooping low under a bridge as they trawled the air for insects.

Yet it was not enough. To understand the river, I needed to access every stretch, to feel its flow and gain an insight into the nature held within its bounds. I was also certain that much of this lower part was the perfect place for beavers to live and was keen to explore further for their signs. The solution was obvious, I needed to take to the water, so I was delighted when local canoe expert Janet Peck invited me along for a wildlife exploration trip from Alva all the way down to the inner Forth at Cambus, where the river spills out into the Forth estuary.

As we set off on a calm October morning, Janet's 15 ft single-paddle open Canadian canoe glided effortlessly ahead while mine zig-zagged erratically as I tried to come to terms with how to guide it properly. My paddle splashed and floundered, whereas hers gracefully swept the water with barely a ripple, so easy and true, and a sign of many years of canoeing experience.

While Janet could maintain a straight course by just paddling from one side of the boat by means of a deft little twist each time the paddle reached the end-stroke, I found the only way I could maintain a clean line was to alternate the paddle from side-to-side: two strokes one side, and then up and over, and two strokes the other. Each time I did this, Janet, who was ahead of me, looked round in time to catch me doing the unacceptable. Despite her coaching, the correct technique was proving a challenge for me to master. Thankfully her patience held firm and she eventually let me be – at least I was moving forward.

The most striking element of this part of the river, even just after we had departed from Alva, was how slow-flowing and languid the waters were – more canal-like with the water looking dark and brooding, which was in stark contrast to the upper part of the river, which is fast flowing and tumbling.

Not long after we had set out, a blur of dazzling cobalt whizzed low over the water ahead of us and alighted on a willow – a kingfisher! We stopped paddling, letting the current draw us forward, but kingfishers are notoriously shy, and this one quickly took flight and shot away out of sight. Kingfishers are on the edge of their range in this part of Scotland and are vulnerable to cold winters. But they seem to be doing well on the Devon now and I had seen them regularly on the river over the past few months.

Kingfisher

I did, however, wonder what impact the unusually wet summer just past had wreaked upon the local kingfishers. I imagine that just as so many sand martins had succumbed to the rising water levels, some of the river's bank-burrowing kingfisher nests would have suffered the same fate.

A line of bubbles appeared ahead of us, and a dabchick bobbed up in the water. It seemed unsure what to make of the two canoes, before deciding upon safety first and diving under the surface once more. Dabchicks are the masters of concealment, and I knew from watching them on the river before, that this bird would have swum underwater to the bankside, from where it would hide in the thick overhanging vegetation waiting for the potential threat to pass.

Dabchick

I tend to only see dabchicks, which are also known as little grebes, on the river in autumn and winter, rather than during the breeding season. This is probably because they build a floating platform as a nest that is anchored to submerged vegetation; totally unsuitable to the vagaries of the regular rise and fall of the river throughout the year. Besides, I suspect predation of their easily found nests by mink and otters makes riverside nesting a rather precarious thing to do.

As we canoed, crack willows and goat and pussy willows arched over many parts of the river, verdant in their growth and in some ways reflecting a rich and luxuriant Amazonian rainforest type environment, such was the manner they encroached upon the water's edge.

Crack willows are so-called because the speed of growth of the trunk means that the bark often splits open, sometimes creating weird marbled patterns beneath. These trees are compulsive crackers and tumblers, and from the fallen trunks, new willow poles miraculously sprout, pointing heavenwards. The crevices in the gnarled bark on the main body of the tree provide valuable shelter for invertebrates and nesting places for wrens and other birds and their roots help to bind and maintain the stability of riverbanks. Willows are thus one of the very bedrocks of the river.

We paddled further downstream, and as we did so, one of the most striking elements was the abundance of Himalayan balsam on the banksides. The plant was everywhere, thick stands carpeting the ground close to the water. The extent of its prolificacy had not really been brought home to me before, even from the large swathes I had found so recently when on foot near Menstrie.

There were other non-natives, too, most notably Japanese knotweed, with its dense bamboo-like growth dominating a few large areas of ground, but nowhere near to the same extent as the Himalayan balsam. In winter, Japanese knotweed dies back to ground level then by early summer the bamboo-like stems emerge from rhizomes deep underground to shoot to over seven feet, suppressing all other plant growth.

I recall as a child that their long rigid hollow canes made excellent pretend swords, from which my brother and I could fight and duel, although an untimely thwack across the body was painful, such was the hardness of the stems. They can spread from even the smallest segment of root, and thus, rivers are the perfect conduit from spreading the plant to new areas downstream.

At least the knotweed and balsam do not reach the full intensity of their growth until later in the season, which enables some bankside native spring plants such as wood anemones and ramsons the opportunity to emerge into full leaf and flower, before wilting back once more, ready to repeat the cycle the following season.

A short while later Janet spotted a problem ahead of us: the whole river here was blocked by a dam of fallen trees and twisted branches, creating what seemed like an impenetrable barrier from one bank to the other. She warned that we might have to drag the canoes out of the water and around the obstruction. This was not an appealing thought, as both banksides were steep and covered in a thick growth of balsam.

We inched slowly towards the dam, which in canoe parlance is called a 'strainer', and which had been created by a large fallen tree trunk having become jammed horizontally from bank to bank, and which had then been backed up by a procession of other smaller tumbled trees and branches over the course of time. A ray of hope; there was a small gap in the dam by the far bank that our canoes could probably squeeze through. Janet went first and made it straight through, thanks to some skilful paddle-work, whilst I grounded in her wake on a partially submerged part of a tree trunk. I was stuck and paddling was getting me nowhere, so I gripped onto a branch above me and pulled my canoe over the obstruction and out into the open water beyond. We were lucky, for the river was high from recent rain, raising the water level just enough for us to negotiate the submerged branches.

In among this tangled wooden jumble on the far side of this natural dam, numerous plastic bottles and junk bobbed in the water; a sad reminder of how rivers act as pathways for our rubbish, carrying it out to sea where our oceans are becoming inundated with both large plastics and micro-plastics. Typically, approximately 5,000 items of marine plastic pollution are found per mile of beach in the UK, and it is thought that more than eight million tonnes of plastic enter the world oceans each year. There may now be more than five trillion macro and microplastic pieces floating in our seas.

It is one of the biggest man-made calamities of our times, and despite all the words of good intention in bringing it to a halt, this plastic pollution perversion simply gathers pace. This is not just a problem for governments and manufacturers to tackle, but one for which every one of us must take responsibility by choosing carefully what we buy and how we deal with it thereafter. If consumer pressure makes it unacceptable for products to be heavily plastic packaged, then manufacturers will take heed and change their practices.

I wondered whether this natural dam might impede salmon, sea trout, eels and river lampreys from migrating upstream. I suspected not, for most of the obstruction was most likely created by buoyant tree trunks and branches on the surface, and there would be plenty space below for fish to negotiate. From my previous snorkelling on the river, I also knew that such underwater entanglements of branches and tree trunks were important for wildlife, providing perfect places for fish and invertebrates to hide and forage for food.

Fortunately, a river never stands still, and one day in the future, when the water is high and flowing strong, this barrier would most likely implode from the gathering pressure behind it, sending a spectacular raft of wood tumbling down the river, and perhaps creating a new strainer lower down.

Our canoes bobbed away from the dam and we gently glided downstream once more. Soon, we came upon several mute swans in the river about a couple of hundred yards ahead. Janet urged caution as swans can be unpredictable in how they behave. We nudged slowly along the left bank, giving the swans a wide berth should we catch up with them, but instead they began to run along the water towards us, taking to the air and swooping low over our heads on whistling wings. It was a quite magnificent sight, a reminder of the power of nature and its grace and elegance.

As we passed Menstrie, the river veered southwards for its last stretch past the edge of Tullibody and then down to Cambus. There were noticeably fewer trees by the bankside on this final leg, with open farmland running straight down to the water's edge. On the more heavily wooded areas further upstream, we had not spotted any tell-tale signs of beavers. That did not mean they were not there, as often their felling is carried out on dry land several feet away from the bank edge, and we could easily have missed their signs. It was nonetheless disappointing, as I was convinced that we would have found some indication of their presence, especially since many parts of the river here had looked perfect for them, being quiet and slow-flowing, with an abundance of willows, alders and other trees.

Willow

It confirmed my suspicions that beaver numbers on the Devon were undoubtedly low, perhaps a handful of animals at most, and that colonisation was likely to prove a slow process. I also momentarily speculated whether the beaver signs I had previously found further upstream could have been made by only the one animal, and that there were no more creatures on the river. But I quickly brushed such a possibility aside, for I had seen indications of their presence on other parts of the Devon over a period of several years. Besides, a recent Scottish Natural Heritage report had indicated there were signs of two beaver territories on the Devon. A beaver territory typically consists of a family group, ranging from two to seven individuals, although individuals may also have non-breeding ranges.

As our canoes neared Cambus, the muddy smell of the estuary drifted across the breeze, and large trees once more swept along the banksides, including a most magnificent horse-chestnut with its large autumnal lobed leaves glowing like ochre. Janet hitched a tow rope to my canoe, mindful that my inexperience might lead to my craft tumbling down the large weir that marks the top of the Devon estuary.

As we pulled our canoes out of the water by the edge of the weir, we both reflected upon the remoteness of this lower reach of the river from Alva to Cambus, where we had hardly seen a soul in our three hours of paddling, despite it lying in a populated part of central Scotland. By taking to the water, we had experienced a wonderful tranquil wildlife world; a place where surprises had lain at every turn.

Chapter 8
MORE ENCOUNTERS WITH BEAVERS
(October)

Mindful from my canoeing uncertainty over the status of beavers on the river, in October I investigated once more the area of the lower river where I had come across their signs previously. It did not take long to find the freshly gnawed stumps of a small alder tree on the top of a grassy bank, which had been coppiced in a most effective manner. This was a remote section of the river, not much frequented by people, so I set my trail camera and left it for a fortnight.

On my return, I could see that the gnawed alders had not been touched any further from before, so it was without enthusiasm that I examined the contents of the camera. While there was no film of beavers, many other video clips had been taken, which underlined the abundance of river life around, including roe deer, moorhens and foraging magpies. There were also a couple of wonderful daytime clips of a pair of otters, cavorting together on the bankside, one of them rising onto its hind legs to get a better view of the surroundings. They were almost certainly a mother and her full-grown cub.

Since June, I had been walking a section of the middle river on an almost daily basis at dawn in the hope of spotting otters, but to no avail. I have, however, over the years seen otters several times on

Magpies

the river, mostly when fly-fishing in the evening. On each of these occasions, I have always been surprised at how unconcerned the animals are by my near presence.

One time, I was atop the riverbank looking down at an otter right below my feet swirling around in a shallow, calm, backwash pool. The otter twisted and turned like a dervish, corkscrewing its body continually, which gave me the impression that it was deliberately stirring up the silty and fine-gravelled bottom. Was it doing this to attract minnows and small trout to feed upon the resultant detritus, which it could then snap up? Probably. Otters are intelligent animals and will have learned this behaviour by accident, or by watching their mothers. Certainly, when I wade in the river when angling, small fish often dart around my feet to feast on the titbits revealed as I tread carefully over the shallow riverbed.

One otter encounter I recall was also in the middle river, this time a pair in the water on the far bank from where I was fishing. Again, the animals paid little attention to my presence, more interested in twisting and turning in among the roots of bankside alders. But such sightings are rare, and unlike our coastal otters on the west coast and in the Northern Isles, river otters are mainly nocturnal and elusive.

With the days getting shorter and the nights longer, I realised that my best chance to spot an otter would be the coming spring. But I would keep my dawn vigil walks going over the winter, for otters have the capacity to appear when you least expect it.

Like the otters, the beavers were proving mighty frustrating, too, with my only success being that brief film of one's back taken by my trail camera back in the summer. Then a friend tipped me off about fresh beaver signs near where I had previously set my camera. I sought out the spot, and it looked perfect – two willows, not yet fallen but heavily gnawed around the lower trunk. One of the willows stood precariously on the narrowest sliver of wood. A beaver would be back to finish the job, of that I was certain.

Night-time watching would be unproductive as I would see nothing in the long hours of autumnal darkness, so I set my trail camera once more in the hope of catching a beaver on film in the act of gnawing one of the trunks. Mindful that my scent might spook them, I returned a few days later to inspect the willow from a distance to see if it had been tumbled, but it still stood proud and tall and I could not see any obvious signs of renewed beaver activity. This had all the hallmarks of being another frustrating encounter with these elusive animals.

When I finally retrieved the camera a week later, the willow was still standing. I was not hopeful at all by this stage, but nonetheless examined the footage on the trail camera later that evening. A lot of videos had been taken, however as I scrolled my way through them, most of them had been activated by the precarious beaver tree swaying in the wind. The days of this willow were clearly numbered, but once fallen, it would become the focus of new life, a place for fungi and wood-boring insects to prosper within its decaying embrace.

I continued my camera examination, and then a clip of a beaver materialised, mostly obscured by an adjacent willow. Then some more beaver film, including a series of amazingly clear night-time infra-red shots of it gnawing and feeding on the tree. Wonderful, this was the break-through I was looking for and I watched spellbound at this most incredible animal at work.

The most striking element was its bare grey-skinned paddle-tail, quite unlike any other mammal that lives upon our shores and a special adaptation for its aquatic world. The beaver would rise up on its hind legs, grip the trunk with its forepaws and begin gnawing, not always at the narrowest bit of the trunk, but stripping off layers in a broad band above and below, often twisting its body to get better purchase.

On several occasions, it paused to eat the bark and wood scrapings it had peeled off with its large teeth, sitting on its haunches and holding the item in its forepaws and feeding in much the same way as a squirrel would with a nut. The animal appeared only the once during the whole week, with the first clip taken at 8.15pm and the last at 8.40pm. This animal certainly was not a habitual returner, and it seems that beavers can wander far and wide over their territory during a night.

Beavers fell trees for a variety of reasons, sometimes to make lodges and dams, but also to access their leaves, bark and twigs for food. I've seen no evidence of beavers building lodges with branches on the Devon, the animals preferring to burrow into banks for their homes, with the entrances often concealed under the water.

Later that week, I ventured a mile or so upstream from where I had captured the beaver film and found more signs: a felled tree and with one of the branches completely stripped of bark. Was this a separate beaver territory? I liked to think that it was. Whatever the case, after the lack of evidence of their presence on some of the other lower reaches, I was once more content that beavers had, at the very least, a tenuous foothold on the river, with every likelihood that they would spread further over the coming years.

Chapter 9
THE MIRACLE OF SALMON
(October–November)

As autumn took hold, I was aware that salmon were now on the move, rapidly ascending the river to spawn on fast-flowing and well-oxygenated gravel beds. Over this period, and into December too, I ventured down to the weir on the middle river as often as possible in anticipation of seeing these magnificent fish jump the white-frothed falls. In the early days there was no sign of salmon, but sea trout were making their presence felt, and I watched enthralled as fish, weighing perhaps two or three pounds, flung themselves at the weir.

There are many natural enigmas: the eel is one and the trout is right there at the top, too. As I had already discovered, unique populations of diminutive brown trout prosper in nearby hill burns, and in the main river they are ubiquitous, haunting every single part with many growing to a good size. They are both predator and prey – a keystone species – and without trout, the river's very soul would be much diminished.

The sea trout is just one element to this rich tapestry of trout variation that adds to their overall mystique. It is of the same species as the freshwater brown trout but is migratory, spending much of its time in the sea, fattening-up on the rich marine feeding and returning to our rivers in the summer and autumn to spawn. I recall seeing sea trout when snorkelling in the sea off the Berwickshire coast many years ago, and they are coastal species in their marine stage, perhaps even favouring the shallow inter-tidal zone, and not travelling nearly the same large distances as salmon.

I imagine the Devon's sea trout spend most of their lives in the inner Forth estuary, and perhaps beyond into coastal regions of the North Sea, where a wide diversity of food, including crustaceans, molluscs and small fish such as sprats and juvenile herring will spur their growth at a much faster rate than their river-living cousins. When they first begin to run the river in summer and when fresh in from the sea, they are magnificent to the eye, silver-flanked and oozing with power.

One of the best wildlife books I have ever read was not a nature book at all, but rather one on how to fly-fish for sea trout. Written by Hugh Falkus, *Sea Trout Fishing* (H F & G Witherby, 1962) is a marvellous and comprehensive study of the sea trout and its habits – a fish for which the author had huge respect. Falkus contended that sea trout were unpredictable and often fickle in their behaviour after entering a river from the sea:

> *Some sea trout stay in a pool for only a few days before running further up-river. Others remain in the same pool for most of the summer... From observations of marked fish that have reappeared in the same places after a succession of spates, I am of the opinion that every pool has its resident population, most if which 'belong' to that pool for the greater part of the season – moving out only when spawning time approaches – the numbers varying from day to day or week to week as other fish move in and rest awhile before continuing onwards to their destined pools.*

Such observations are borne out by experienced anglers on the Devon who have told me that sea trout can sometimes be seen entering the lower reaches of the river in the summer, but then seem to lie low in pools before starting to run the river properly in September once their skin has become coloured.

By the middle of October, I began to see the first salmon trying to negotiate the weir and the temptation to take the fly-rod and try my luck at catching one proved irresistible. Fishing is not permitted by the weir because that would be unfair to the fish, so early one evening in October, and after a period of rain but with the water still running clear, I cast a fly across the river about a mile or so further down-stream, letting the current carry it naturally in the flow. Fly-fishing is an absorbing pastime and a passion for many.

Salmon

Concentrating on the movement of the fly and waiting for the pull of a fish is compelling and empties the mind of all thoughts. It is the perfect relaxant. I'm no expert angler, and while I once hooked and lost a salmon, I have never been lucky, or skilful enough, to land one. At one time that bothered me, but not any longer, as for me, size does not matter when fly-fishing, for the satisfaction of flicking a fly towards a rising ripple and have a small trout take the lure is as exciting an experience as any I can imagine. Nowadays, I always fish with barbless hooks. More trout are lost that way, but it is better for the fish, and for me, that is important.

As ever, I caught no salmon that evening, nonetheless I enjoyed being out on the river, watching the soft autumnal leaves being carried down in the water flow and experiencing the magic of a large mixed flock of rooks and jackdaws tumbling in the evening sky and calling excitedly as they headed to their night-time roost on a nearby wooded ridge.

Finding it impossible to get salmon out of my mind, I returned to the weir the next day to see if there was any more activity. The weir is a formidable inclined waterfall, about six feet in height with a fish ladder in the middle that had recently been refurbished in a bid to help passage.

I watched for a while, the cascading water almost hypnotic in quality, but there were no fish. Then, the action began as the first of several flapping salmon threw themselves at the foaming white torrent. Some made it half-way up, their tails pumping furiously in desperation to overcome that last few feet to the top. It was, sadly, always a case of 'so near, so far', and each salmon failed the challenge, slipping slowly back into the churning cauldron below.

The sheer willpower that drove these fish forward was both inspiring and humbling. For every failed attempt, a little bit of precious energy was lost and with it the chance to successfully spawn. Each time a salmon leapt into the air, I willed it to succeed, but the river was winning, always winning. It was like watching an unfolding tragedy. These fish had already been through so much, surviving in the river as young parr, then heading out to the open sea where countless other dangers lay, before returning to the river once more.

More fish jumped, but the rapids continued to beat them back. These salmon were the very epitome of grit and determination, their hormones driving them onwards to their upriver breeding grounds. Even those lucky ones that do manage to spawn will most likely succumb afterwards, every vestige of energy having been sapped from their spent bodies.

Despite the prolificacy of failed attempts, many must make it over the top, and later in the week I witnessed a large salmon jump clear of the water in the gentle glide above the weir, almost as if in celebration of its achievement. Besides, my previous electro-fishing experience in the Dollar Burn upstream of the weir had revealed young salmon, the progeny of the previous year's spawning, and a clear sign that a good number of salmon beat the challenge of the falls. It is one of nature's miracles.

A few years back, I had visited the same weir at night-time under the soft illumination of the moon, for I was curious whether the fish were more active at night when they would presumably feel safer from predation. It was a wondrous experience to sit by the roaring water and I could see instantly that the gentle glow of the moonlight combined with the gurgling whiteness of the falls would make it easy to spot any leaping fish. A heron that had been standing by the edge of the tumbling water ghosted away into the night – an encouraging sign that fish were on the move.

Seconds later, a fine salmon grilse of around four pounds in weight slapped against the side of the weir. Several more soon followed and it was apparent that a much larger movement of fish was taking place under the cloak of darkness than typically occurs during the day.

I was also intrigued to glimpse some very small fish, perhaps no more than six inches in length attempting to jump the falls – and no doubt the reason why the heron had stationed itself by the edge of the water. Despite their small size, I can only think that these were sexually mature brown trout that were also consumed with the urge to spawn.

For such a large fish, it is quite astonishing how far salmon will go up the headwaters to spawn, often in places where the river and associated streams are narrow and very shallow. I recall when hiking in the Cairngorms, or in the Angus glens, coming across salmon in such high burns in autumn.

Unlike some Scottish rivers where salmon might start moving upstream as early as February or March, salmon in the Devon are very late runners, usually not appearing in the river until September before moving quickly up to the higher reaches and the lower parts of the tributary burns and spawning in late November and December. Most salmon die after spawning, their bodies spent from the effort of running the river and breeding. I do wonder, however, whether survival rates are better than average on the Devon, as the home run back downstream to the sea is relatively short.

I've occasionally come across dead adult salmon in early winter lying by the bankside of the Devon. When I lived in Aberdeenshire, they were a frequent sight along the Dee and its tributaries such as the Water of Feugh. Indeed, great black-backed gulls on the Dee and its tributaries had developed the habit of scavenging for the bodies of spent salmon (kelts) that have just spawned, and the sight of one of these large imposing black and white birds swooping between the alders by the banks of a gushing river is as great a natural spectacle as any I have ever witnessed.

I was keen to catch some salmon in the act of spawning, so for a period of several weeks that autumn and early winter I scoured likely-looking stretches of the Devon where the water flow was fast and the riverbed pebbly. It is sometimes possible to detect their presence by the bow waves produced as they work their way through the shallows. Salmon are skittish fish, and I always trod carefully along the banksides as I had no wish to disturb any fish when spawning.

Despite persistent searching, I was not lucky enough to find any fish in the act of breeding due to the persistently high-water levels. I did discover, however, a spawning bed several hundred yards above the weir – known as a redd – where the hen salmon had excavated a furrow with her tail to lay her eggs, which are then fertilised by a cock fish before being covered up with gravel and pebbles. It was hard to discern the redd at first under the rushing water, but a glimmering of paleness on the gravel bed was enough to merit further examination with my binoculars. On closer scrutiny, it was clear that the gravel had been disturbed and turned over, forming an oval-shaped patch about three feet in width, which also looked marginally raised, like a shallow dome. The amount of energy required in making this spawning bed must have been considerable, and I could not help but wonder at the abrasive skin damage the hen salmon must have endured during the whole process, making her highly vulnerable to infection and disease.

For the salmon, its raison d'être had been fulfilled, and while she had expended a huge amount of energy, the next generation was lying as eggs buried beneath pebbles on this riverbed ready to emerge as tiny alevins (fry) in a few months' time. It was like the beating of an invisible heart.

Not long after discovering this redd, the winter was characterised by unusually heavy rainfall, sending the river into tumultuous flow on several occasions. Such increased incidences of high rainfall could well be one symptom of climate change, and the resultant spates have the potential to devastate salmon spawning beds.

Salmon numbers on the Devon are in an uncertain state, and accordingly, it is deemed a 'Category 3 Salmon River', which means because of the fragility of the stock, all fish caught must be returned back to water unharmed so as to aid their conservation. Despite this, anglers that autumn reported much better salmon catches than in the previous year, with around 20 recorded fish caught, compared with only two the previous year. However, this was most likely due to the unusually wet autumn, which raised the river and suited the fish well for their upstream passage, as opposed to any recovery in the stock.

Salmon across Scotland are in decline, and there are a variety of factors behind this fall. Changes in the ocean ecosystem on their marine feeding grounds off Greenland and the Faroes are thought to be the main cause of the decline, most probably because of climate change. Some Scottish fish travel as far as the Davis Strait between the western coast of Greenland and Baffin Island in the Canadian Arctic.

Although there are the deniers, human-induced global warming is an irrefutable fact, and one that was brought home forcibly to me the previous year when I attended a science conference in Aberdeen, which explored the marine impacts caused by climate change. The evidence from the speakers, all of them experts in their field, was compelling; our warming seas are having a real and discernible impact on the distribution of many marine species. In British waters, this is resulting in fish such as cod moving northwards, while warmer water fish such as red mullet are coming in from the south, arriving in areas where they were previously scarce.

What is more, if our marine ecosystem becomes unbalanced, then we end up pointing the gun barrel at the head of humankind, too, given that seafood is the primary source of animal protein for an estimated one billion people around the world. Thus, behind the fall in salmon numbers on the Devon lies a much bigger story, the consequences of which should terrify us all.

Chapter 10
A REMOTE AND WILD PLACE
(October–November)

One of the wildest and remotest stretches of the river is the first part of the middle section from below Rumbling Bridge down to Vicar's Bridge, a steep wooded gorge in places, where the water tumbles through surging pools and over rock ledges. Known as the Devon Gorge, the river here marks the county line between Clackmannanshire and Kinross-shire, and in appearance is somewhat akin to a Highland Perthshire river.

It is a place of mystery and intrigue, and somewhere I deliberately put off investigating until late in the autumn, as I wanted the trees to be largely devoid of leaves, so that enhanced light filtered down to the woodland floor.

The jewel in the crown is the Cauldron Linn, a magnificent two-tiered 40 ft waterfall that cascades down a narrow canyon. *The Scottish Tourist and Itinerary* (Edinburgh: Stirling & Kenney, 1830) – a guidebook popular in the nineteenth century – described the falls here as *"the greatest natural curiosity, and certainly one of the most sublime objects in Scotland"*.

Such a description still holds true to this day, and if the Linn was accessible by car, it would almost certainly be a leading Scottish tourist attraction. It has a dark and imposing aura, heavily wooded all around with steep slopes and rocky outcrops on either side.

There is no easy route to the falls, for every way is rough going. The route from Powmill in Kinross-shire is perhaps the least demanding. However, I wanted to experience the river and its wildlife, so I ventured up from Vicar's Bridge, a walk of about one-and-a-half miles upstream towards the Cauldron Linn, which because of the roughness of the ground takes about an hour.

Vicar's Bridge is a haunting spot, dark and shaded in a rocky cleft with the Devon slowing to a deep dark pool before spilling on its way again just past the bridge. On the bridge is an eroded stone engraving, which reads:

Sacred to the memory of Thomas Forrest the worthy
Vicar of Dollar, who among other acts of benevolence, built this bridge.
He died a martyr A.D. 1538.

69

As I made my way upstream from Vicar's Bridge, the wildness of the woodland shone out at me – an eclectic mix of beech, birch, oak, sycamore, and several other species. Rotten tree stumps abounded, perfect places for great-spotted woodpeckers to forage and excavate their nest holes, while birch polypores, a strange-looking bracket fungus adorned veteran birch trees.

Several of the woodland stretches here hold a proliferation of ash, which is a tree I've always found to be rather peculiar; it's one of the last trees to unfurl into full leaf – usually by the end of May – yet is among the first to shed its foliage in autumn. This would seem to give it a short growing season and put it at a competitive disadvantage with other trees. But this casual observation does not reflect reality, for the ash is prolific in growth and generally abundant.

The late emergence of ash leaves is ideal for sun-loving forest flowers such as wood anemones, lesser celandines and dog violets. And, even once the canopy is fully verdant, the airy arrangement of the leaves enables plenty of light to penetrate below.

It is, however, autumn when I like this ash woodland best, not for its fiery leaf colour (ash leaves often fall to the ground with hardly a turn of hue) but rather because of the special ambience here, in particular the striking pale trunks, which radiate an enveloping soft luminescence under the weak autumnal sun.

While there may have been sunshine in the sky above, I could

Lesser Celandine

not help but reflect upon the dark cloud that hung over the future of this wood in the shape of ash dieback, a recently discovered fungal disease that has the potential to decimate our ash populations. Certainly, some of the leaves on the ground had crinkled dark edges – one of the signs of disease. In this instance, I did not have the expertise to determine whether this was a normal part of autumn leaf fall or was indeed actual infection.

Does the possible catastrophic loss of the ash matter? Well, yes it does, because the tree provides home for a large variety of invertebrates, fungi, lichens and mosses, some of which are highly dependent on ash.

As I continued my way, I put such thoughts of this potential arboreal catastrophe to the back of my mind, much better to enjoy the moment and the wonderful wild woodland and rushing river. I moved down the slope towards the bubbling river and sat for a while by a little waterfall. It was such a tranquil place. A busy wren flickered through the twisted tree roots of a fallen tree and wind-tumbled rooks swirled high above the gently swaying skeletal branches. Down here at ground level, there was hardly a breath of wind, such was the shelter afforded by the deeply incised ground.

Red Squirrel

I struck upriver once more, disturbing a red squirrel on my way, which scuttled up a tree with an acorn in its mouth, before pausing and turning around with its head pointing down the trunk from where it could get a better view of me. With its bushy tail nervously flicking from side-to-side, I managed to snap a quick series of photos, before its nerves got the better of it, and it fled up to the highest part of the tree and disappeared.

A short while later, another red squirrel – busily collecting and burrowing nuts – loomed into view, making the most of this season of plenty. They seemed to be prospering here, which was encouraging as this woodland is broadleaved in the main, the type of habitat where grey squirrels tend to do better and out-compete reds. In recent years, I've noticed reds being more frequent than greys in other parts of this region, stretching from Dollar to Muckhart and right up into Glendevon. When I first moved to the area more than 12 years ago, my impression then was that the grey squirrel was predominant.

If these observations on squirrel dynamics are correct, then a possible reason for this change in fortune is because pine martens have become established in this part of the country in recent years. Martens are proficient climbers and powerful enough to take squirrels. However, the smaller red is so agile that it can usually dodge a pine marten by fleeing to flimsy branches on the very edge of the tree canopy before leaping over to the next tree. Interesting recent research has also shown that red squirrels are better at detecting pine marten scent than greys and are better able to avoid them.

For a pine marten, the less nimble grey is much easier to catch, especially since it spends much more of its time on the ground than the red. Thus, the corresponding reduction in grey squirrels has benefited the reds in terms of reduced competition.

Soon, I reached the top of a steep slope overlooking the tributary Gairney Burn, which I descended by means of an undignified backside slither down to the water's edge. This is a delightful burn which courses its way down from the small community of Powmill. Quite wide in places, I suspected it brimmed with invertebrates, and I could not resist turning over stones and small rocks in search of stonefly and caddisfly larvae close to where the burn joins the Devon.

There were no obvious stepping-stones in the burn, so after my invertebrate search, I waded across, and then rounded a steep spur by the main river to be suddenly hit by the thunderous roar of the Cauldron Linn filling the air, delivering tantalising glimpses of its tumbling water through the trees ahead of me.

The area around the falls is strewn with large moss-covered boulders, slippery and treacherous, and real ankle breakers should a foot inadvertently slip between a gap. I surveyed the scene for a while, trying to determine a safe route before making my final approach. After a bit of cautious clambering, I soon found a suitable vantage point on a large rocky slab and focused my attention on the falls.

It was a most spectacular sight, spilling down over two large vertical stages, with the lower section forming a frothy white torrent that spilled into a large pool below. Totally unpassable for fish, this is the highest part of the river that any migratory salmon can reach, and I imagine that every autumn a few make it as far as the Gairney Burn a couple of hundred yards downstream to spawn.

As I stood by the Cauldron Linn, pondering the natural secrets that lay within its hold, my eyes were drawn to the dark angry pool below the tumbling waterfall. The name 'Devon' almost certainly originates from the Gaelic word 'Duibhe', which means black, and by looking at the pool one could instantly see why the river could be so-called. Dubh-Abhainn translates as 'black or dark river' and prior to the mid-1800s, the river was widely known as the 'Dovan', before the adoption of the current name through the evolution of language over the generations.

However, the *Statistical Account of Scotland (1791-1799)* indicates it is derived from the old word 'Dobh-an' or 'Dovan', which it claimed translates to a swelling or raging water – again, an entirely appropriate description of the river.

I spotted a dark twisted sliver atop a small boulder beneath me, so I climbed down and recognised it immediately as a spraint (an otter dropping), which are often deposited in prominent positions to act as territorial markers. I wondered whether the otter had been fishing in the pool below the falls, perhaps migratory fish had congregated there unable to go any further, thus making it a productive place to hunt.

All around where the spraint lay, a sweeping covering of moss clung to the rocks like a green velvet carpet, thriving in the shaded humid environment create by the gorge and the waterfall. I had a magnifying glass in my backpack, which I quickly brought into play, and the green omnipresence of these mosses took on an exhilarating new dimension when examined closely under the lens, exhibiting an intricate beauty that was a sight to behold.

Moss is very important to our environment as it forms vital shelter, humidity and safe breeding places for a huge array of tiny creatures. These little bugs are the food for a host of larger animals, which means that without mosses the whole food web would suffer immeasurably.

Mosses also protect the ground from erosion caused by wind and rain. Indeed, the very presence of moss can spark new life into exposed and infertile terrain by acting as the catalyst that initiates soil formation. Such instances are found in the woodland in which I stood, for in spring, white-petalled wood sorrel sprinkles these moss-covered rocks and fallen tree branches like scattered confetti, providing a fragile toe-hold for their roots, which although thin, provides enough sustenance for the plants to absorb nutrients and water.

Sadly, it was time to go, as dusk would settle across the gorge in an hour or so, after which time the wood would become a difficult and treacherous place to walk. Later, as I neared Vicar's Bridge, a tawny owl hooted, a quavering 'hu...hu-hoooo'. The sun might still have been just above the horizon, but the dark embrace of the wood had stirred this bird of the night into calling early.

Chapter 11
DEATH FOR SOME,
OPPORTUNITIES FOR OTHERS
(November)

In early November, the first of the heavy autumnal rains deluged the middle part of the river in Strathdevon; starting at night-time, continuing through the next day before easing off the following evening. I knew the Devon would be rising rapidly because the reservoirs upstream were already brimming from the unusually wet summer, and with the leaves now falling and vegetation dying back, there was little natural growth to impede the rainwater and slow its progress into the river.

Late in the afternoon, and with angry grey skies already darkening in anticipation of dusk, I ventured down to the haugh (flood meadow). Stretching for almost two miles down either side of this part of the river and bounded by the Ochils on one side and by the low-lying wooded Sheardale Ridge on the other, this is a truly wondrous place of vast horizons and boggy pools that provide rich feeding and shelter for waterfowl.

As soon as I arrived, it was immediately apparent that the river had burst its banks, spilling over into the haugh and creating a vast inundation of muddy water that swirled around the trunks of willows and alders. It looked much the same way a mangrove swamp might look in more tropical climes. There the similarity ended, for the air was cold and the persistent drizzle blustered about in the wind, misting my glasses and dribbling water down my neck in a frigid trickle.

I approached the edge of the spreading flood as it crept forward with the sureness of an incoming tide, gradually filling cattle hoofprints pock-marked upon the ground. A movement caught my eye by the water margin – a diminutive brown furry bundle, swimming with whirring legs in a purposeful manner towards a small raised clump of grass. It was a field vole; its body buoyant and bobbing like a cork, the fur on the upper parts surprisingly dry despite its unexpected dunking.

Sudden floods such as this are a disaster for voles. They are abundant on the haugh and this creature was no doubt one of many that were trying to escape the rising waters. The vole disappeared into a protruding tuft of grass, which by now was a tiny islet. I waited for a few more minutes, curious to see whether the creature would reappear, but it never did, preferring instead to lie low in its fragile shelter. If the water rose further again, then this frightened and stressed animal would once more have to take the plunge to find a new area of dry ground.

Field Vole

One animal's calamity is another's fortune, and several herons, crows and magpies had gathered by the water's edge on the look-out for these fleeing voles, as well as invertebrates such as worms emerging to the surface, as they often do when the ground becomes saturated. Numerous black-headed gulls were also excitedly flying low over the flood plain, taking advantage of this new-found bounty caused by the spreading river.

Black-headed Gulls

Mole

Goodness knows what happens to moles at such times. They are certainly frequent on the haugh – but are they fleet of foot enough to escape the rising flood waters, or can they survive in air pockets in their elaborate burrow systems? I imagine it must be the latter, although many animals must also succumb.

While some birds were benefiting from this sudden food bonanza, others such as goosanders and kingfishers had disappeared altogether. Despite both species being regular inhabitants of this stretch of river, the sheer strength of the current and the murky water had made it impossible for them to hunt fish and I suspect nearby lochs, or the coast might prove a more attractive option during such periods.

Indeed, kingfishers have a second strategy, which I've noticed on previous occasions when the water is running high, where they move to the numerous backwashes, creeks and ditches that feed into the main river. The water remains remarkably clear in such places, even after heavy rain, and small fish are abundant too because they are sheltered from the surging torrents of the main river. Once I had discovered this behaviour, my sightings of kingfishers greatly increased when exploring such quiet backwaters during times of flood.

What happens to fish during times of spate? While many larger fish such as trout and salmon can ride out the torrential flow by hunkering down close to the riverbed and seeking shelter behind rocks and submerged tree trunks, other fish face more of a challenge. Indeed, after one flood event, I found numerous tiny young sticklebacks and minnows trapped in puddles in the surrounding fields that had been left behind after the floodwaters had subsided.

Darkness was beginning to take hold, so I reluctantly struck for home, aware that the spilling of a river across the land is one of nature's great spectacles. It also highlighted the importance of these flood plains and why we must protect them, for they are natural relief valves that accommodate rising waters, thus protecting our towns and villages.

I returned at dawn the following day. The rain had stopped hours previously, and the river had backed down so that most of the flow was now contained within its banks. However, the haugh was still a watery expanse of pools and puddles, and an intricate oxbow lake nearby was so full that a discernible flow was draining out of it and back into the main river. Oxbow lakes are fascinating landscape features created when a meander in a river becomes cut off from the main flow, leaving behind a crescent shaped body of still water.

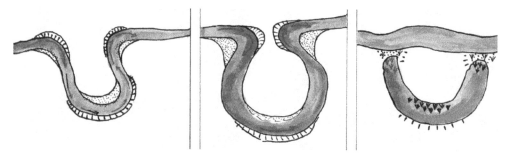

Oxbow Lake Formation

The extent and complexity of this oxbow, which was established many years ago – a natural and distinctive U-shaped channel which is a haven for ducks and snipe – is best appreciated when viewed from above, either from the Ochils or the Sheardale ridge on the other side of the strath.

The previous night's flood had delivered these birds and all the other wildlife that thrive within the oxbow's bounds a significant boost by topping up the water it contains, ensuring it continues to support life for the months to come. Snipe especially benefit from such flood inundations. They are increasingly scarce birds, numbers having dwindled because of land-use drainage, resulting in their preferred areas of habitat decreasing with each passing year.

This is a tragedy, for they are such special birds that exhibit the most peculiar physical attributes. A snipe's eyes are set well back on its head, providing good all-round vision and the elongated bill has a sensory tip, enabling it to feel for worms under the soil. A remarkable facet of the bill is that it can be opened only at the tip if so desired. Thus, working like a pair of intricate forceps, the bird can catch and swallow invertebrates without even needing to withdraw the bill from the ground.

In spring the male snipe has an impressive courtship flight. He flies high in the sky and sweeps round in fast circles, the air rushing through the stiff outer tail feathers creating a most unusual noise, which is often referred to as 'drumming'. In my view it sounds nothing like a drum. I have also heard the noise described as 'winnowing', which is more apt as it does indeed bear some resemblance to a bleating goat.

Most winters I will flush at least one Jack snipe out from rushes by this oxbow lake. A winter visitor to our shores, they are noticeably smaller than our common snipe and sit ever so tight. When I do stumble upon one, it will take to the air in a short flurry before quickly dropping down again. This is in stark contrast to the snipe which when disturbed will disappear into the far distance like a crazy rocket.

The pools and ditches replenished by these autumnal and winter floods form valuable refuges for frogs, toads, newts and a vast range of insects and other invertebrates. Wild plant seeds will also have been dispersed by

the flood waters, bringing colour to the haugh with the pastel-purple blooms of cuckooflower in spring acting as a magnet for early-emerging butterflies such as orange-tips. The downside is that these same waters will be spreading Himalayan balsam seeds to new areas, including areas well away from the riverbank.

Cuckooflower *Orange-tip Butterfly*

As I wandered by the fringes of this oxbow on the morning after that autumn deluge, I inadvertently put to flight a pair of teal. I watched them wheel into the distance, regretful they had been disturbed, although at the same time they did bring warm memories flooding back to the mind. One of my favourite winter experiences is to seek out a secluded vantage point above these rush-filled pools just prior to dusk, from where I can watch teal flying in to settle in their boggy margins to feed. As darkness descends, these wonderful little ducks can be heard whistling to each other in the night air, a soft and flute-like call, a mere whispering in the wind. It is the essence of winter on the haugh, the call so gentle and hypnotic that it is the very music of nature itself.

Teal

Chapter 12
AN OASIS OF WILLOW CARR
(December)

A cold December morning on the middle river, the first creeping embrace of dawn still several minutes away, although high in the eastern sky there was a glimmering, so faint it would be easy to miss, but a herald nonetheless that daybreak was near.

Although still dark, I was used to being out at such times and had engaged in dawn-walking this section of the river most days since June. Back in the summer the walk was pleasant and revealing, the early mornings bathed in brightness and nature at its most frenetic in rearing young and ensuring the continuation of the next generation.

In early winter it was so very different. Cold and dark, the sun still to break the horizon, my eyes tune as best they can to this nocturnal world, with sound as important as sight; my ears straining for the slightest noise or ruffling.

Eyes predominate overwhelmingly in our perception of the world, with our reliance on other senses much diminished. Out on the river when dark, my ears take on a new importance, listening intently to detect the tiniest scuffle in the vegetation, the fluttering of roosting fieldfares in a willow thicket, or the call of oystercatchers up high somewhere in the night sky.

It is like a sensory jigsaw puzzle, piecing together fragments of blurred outlines and subtle noises to gain an overall picture of nature at night. This is how we must have been in prehistoric times, our souls part of the wild landscape, using every sense at our disposal, including smell, in our quest for food and to detect danger.

As I trod carefully along the bank, lest I should slip on the muddy path, there was a loud and sudden splash in the water – a large fish, a salmon perhaps. As quick as the first commotion subsided, the water swirled further, and as I peered down from the bank into the river, more ripples spread out across the surface.

A dark head bobbed up, and then a second – a pair of otters! Finally, after six long months of walking the same river circuit at dawn, I was experiencing my first face-to-face encounter with these special animals. Although daybreak neared, the light was still extremely poor, but just good enough for me to follow their indistinct forms as they made their way slowly upstream, periodically diving and then surfacing again, before disappearing under the roots of a bankside alder. I waited some more in the hope they might reappear, but by using the cover of the alder roots, they had slipped away into the early morning darkness.

Barn Owl

I felt a strange mix of excitement and frustration – disappointment at the encounter being so brief, but thrilled that my perseverance in dawn-walking the river had eventually revealed a sighting, especially since otters were one of my main focuses of these morning ventures. But as each month passed without seeing an otter, they had slipped to the back of my mind because I would invariably stumble upon something of interest.

Only the week before, the pale form of a barn owl had floated past on slow-beating wings, and on that same morning, a fox trotted along the river path towards me, before picking up my scent and fleeing in panic.

I had already decided that otters or no otters, I would continue my early morning outings for the rest of my river year, which would end the following May. Perhaps I would spot more otters during such morning wanderings, but I instinctively knew from this short encounter that to get to know otters better, I required a different approach. Rather than random walking, it would prove more fruitful to seek out an area where otters frequented during the longer daylight hours

Fox

of spring, ideally close to a holt containing young cubs, and then pick a good vantage point to observe them from.

That was all going to be several months away, but surprisingly, otters reared back into my mind only a couple of days later when I retrieved my trail camera from a nearby backwash of the river. On examining the camera, it had captured on film the same pair of otters that I had just seen that dawn morning. These otters were clearly fond of each other and engaged in brief mutual grooming, before disappearing down a small watery channel.

Could these have been the same two otters I had caught on video a few weeks previously in October further downstream? Yes, I imagine they were, for otters wander far and wide, which is one reason why it is so hard to plan to watch them on the river.

This backwash or creek where I had set the otter camera was fast becoming a place of wonderment for me. It is a most unusual physical happening, an illustration of how the river creates habitat and sculpts a new environment from the power and vagaries of the rushing current. Twelve or so years previously, this part of the river was a sharp left-turn bend that flowed relatively freely and true. Then, a flood event had thrown everything askew, creating a shingle island close to the outside turn and developing its own subsidiary channel away from the main river.

Over the following years, the main part of the river changed its course slightly once more, taking it further away from the shingle bank, thus resulting in an expanse of calm water by the original outside bend – a large pond in effect, a backwash where, except during times of flood, the water is always calm and still.

While these complex hydrological processes were taking place, the shingle bank was gradually being colonised by willows and other plants such as broom, monkeyflower and water forget-me-not, much in the same manner as how a recently created volcanic island in the sea soon becomes colonised with life. The shingle bank had also developed two inlet creeks

Broom

inundating deep into it, resulting in a diverse mix of young willow carr and still open water that had become a rich wildlife haven.

When I returned to reset my trail camera not long after this most recent otter video capture, I took some time to explore this little oasis in more depth, criss-crossing my way through the thick willow growth. Although the willows were devoid of leaves because winter had taken full grip, the density of their saplings was quite astonishing, making it difficult to walk through and gain access to one of the creeks. Willows are incredibly fast growing, and in a space of only a few years they had created this jungle-like thicket. The most striking element was the way the density of the willows had acted as a trap, capturing leaves, branches and other natural flotsam carried by the river when in flood, and leaving in its wake a thick carpet of decaying vegetation once the water had subsided.

I scooped my hands into an accumulation of these leaves and already the process of decomposition had begun. In time, this little handful would transform into leaf mould, returning valuable nutrients back to the ground. As I placed the claggy clump back on the ground, a couple of tiny millipedes wriggled across my fingers, another indication of the importance of these fallen leaves as a shelter for invertebrates. During heavy frosts, this leaf litter acts as an insulating blanket, providing a haven for micro-creatures that may otherwise succumb if exposed to the full winter cold. Blackbirds and thrushes forage in such areas, turning over the leaves with quick flicks of their beaks in search of the hidden bounty that lies below.

Millipede

In effect, these willows were creating soil from the trapped rotting detritus, giving new pioneer plants and invertebrates the opportunity to colonise and gain a foothold. It was nature at work, the river at work, creating a new ecosystem of plant and animal communities. I found this totally compelling: there are so many such miraculous natural processes going on around us all the time, but their subtlety can easily be missed, as is their importance to the overall wellbeing of the environment. When we dam or dredge a river, create a weir or extract water, then we are upsetting such natural forces, resulting in reduced natural diversity and a diminished environment. It was a watershed moment for me: I knew rivers were important and special to the environment, but within that clump of decaying leaves I had found one of the fundamental reasons why.

Deep in thought, I reached one of the watery inlets, putting to flight a pair of dabchicks that had been foraging by the waterside. These little calm bays are perfect for dabchicks to forage because the thick sediment below is rich in invertebrates and act as sheltered nursery grounds for young fish. I set up my trail camera overlooking this little inlet, and on retrieving the camera several days later, the series of videos revealed a multifarious selection of birds reliant upon its enveloping calmness. The pair of dabchicks were frequent visitors during the day, almost comical in appearance with their fluffed-up rear-ends, and mallards, too, were abundant. After dark, little teal would swim this way and that across the water, their bills sifting the shallows for food.

Mallards

Each time I approached this shimmering patchwork of water and willows, my heart would begin to race in anticipation as to what might turn up. The previous summer, sedge warblers reared a family from within a lush tangle of brambles, grasses and rosebay willowherb by the water's edge; the cock bird often using the top of one of the nearby willow saplings as a song-post, his head swivelling from side to side to ensure his rather grating musical notes were projected in all directions.

During another summer visit, I sat upon a log embedded in the mud by the river's edge and gazed into the shallows in quiet contemp-

Sedge Warbler

lation and as I did so, a small fish tinged with red and blue darted below. It was too far away to examine in detail with the naked eye, so I rose to my feet and stepped back slightly to bring it into focus through my binoculars.

It was a male three-spined stickleback and what a fine-looking fish, resplendent in full breeding colours with a bright red throat and underbelly, a blue splash on the gill covers and iridescent green back. Although I could not see any structure, somewhere nearby would be his tunnel-like nest constructed out of small plant fibres and bound together by sticky threads.

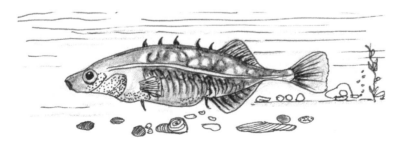

Stickleback

This stickleback had only one thing on his mind – to entice a female into this intricate little cocoon to lay her eggs, which he does by engaging in an elaborate zig-zagging courtship display where he shows off his striking colours like some miniature underwater peacock.

After the eggs have been deposited and fertilised in the nest, the male stickleback will guard them like a hawk, fanning water over the precious clutch with his fins to ensure they are well oxygenated. Once hatched, the plucky male will continue to guard the fry until they have absorbed their yolk sacs, after which they disperse. The whole breeding cycle is as equally miraculous as that of the salmon.

As summer turned to autumn and then to winter, I frequently encountered herons and kingfishers by this little offshoot of the river. A male kingfisher had taken up residence here and he became a familiar sight in autumn and early winter: a piercing whistle followed by a bolt of electric blue as he shot along the course of the river, flying with the surety of a guided missile. Sometimes the sighting was so brief that I had to question whether I had seen anything at all. Could that flash of cobalt have been just a figment of my imagination?

Well, no, the blue-blurred form was always real enough, for that is the way with kingfishers, given that they are such flighty birds, quick to take to the air and whooshing low over the water with verve and determination. Indeed, it is unusual to ever get a close and clear view of a sitting bird, such is their shyness.

Kingfishers are creatures of habit and have their favourite fishing perches, often an overhanging branch from a tumbled tree protruding above the water's surface. The quiet creeks of this part of the river provided the perfect place to hunt for fish such as sticklebacks.

I observed this kingfisher regularly by the same spot right until almost the middle of December, but then winter torrential rains fell on a biblical scale for several days, raising the river to the highest level it had been for quite some time and swamping this formerly quiet area of backwash and willow carr. The silt-laden brown tumbling water was more than any kingfisher could endure, and it had no option other than to move on and seek quieter haunts elsewhere where the water visibility was more conducive to spotting small fish to prey upon.

By this stage, much of the haugh was underwater, including a rush-filled pasture that did not usually flood. Here, a large flock of black-headed gulls and teal had gathered, prospering in this newly created shallow mosaic of pools and bog. It was a marvellous sight, the gulls busily scrabbling their feet on the bottom to stir up invertebrates, while the teal gently whistled to themselves. It was clearly a productive place to feed with the gulls feasting upon an abundance of worms, gobbling them down with enthusiasm.

On another day during these December floods, the unexpected happened. On ghostly wings a large white, long-necked bird floated into view by the field edge, an angry heron trailing in its wake. I was momentarily stunned, not quite believing my eyes, for it felt like I had been carried away to some southern European river delta or marsh. Surely not, I thought. I scrutinised the white bird through my camera's telephoto lens, and it was exactly what my first inclination had told me – a great white egret. Normally found on the Continent and into Asia and Africa, this was a rare and unusual visitor to these parts, its white-cloaked elegance drawing my breath away.

Great White Egret

The heron, which had landed beside the egret, was not so impressed by this foreign interloper and it stabbed its long sharp beak towards it, sending it into the air once more, with the heron taking chase, before both landed again by a watery margin a short distance away. The heron looked set to attack again, but a walker on a nearby farm track inadvertently spooked it, sending it back into the air to settle in a far corner of the field. The egret, on the other hand, was unconcerned by either the walker or by my near presence, and it happily stalked the shallows, before taking to the air in intermittent short flights to seek out new areas to forage. I wondered whether this egret had hailed from a remote marsh or lake in an unpopulated part of Europe, which was why it was so relaxed in the presence of people and unconcerned about any danger. These birds, along with their little egret cousins, have spread northwards dramatically over the last couple of decades from southern Europe – a further example of climate change affecting the distribution of species.

Elated by this most unusual encounter, I struck for home, pulling on my woolly hat to stave off the cold December drizzle that had resumed once more. This was a *river of surprises*, a place of the unexpected, and a place where I craved for more, an entrenched addiction that was proving impossible to quell.

Chapter 13
CALL OF THE WILD GEESE
(December–January)

Dusk on a late December evening by the estuary of the river by the village of Cambus and in the distance, somewhere towards Stirling, there was a murmuring, ever so faint at first, but gradually gaining in intensity.

It was the call of the wild geese, scores of greylags and pink-foots making their way from their feeding grounds in the fields of central Scotland and down towards their roosting sites by the inner Forth estuary. With the cold air edging towards frost, I scanned the gloaming mauve sky in anticipation, eagerly seeking out their V-shaped formations etched against the fading heavens.

Pink-footed Goose

Although their honking was getting louder, there was no sign of the geese at first, but then, over the trees fringing an area of pools and reeds, they materialised, a flock of greylags swooping over in a low arching flight, the air ringing to their excited calls. They spotted me from my vantage point on a small grassy knoll and veered away on powerful wings, before heading across the narrow expanse of the Forth, into which the Devon estuary flows.

That was just the beginning. Further away and much higher up in the air, two large skeins of pink-footed geese swarmed across the sky, flying with real purpose and direction, their calls higher pitched and more metallic than the greylags. I wondered whether these birds had come from Iceland or Greenland and conjured in my imagination the wild open scenery that must unfurl across their breeding grounds – places backdropped by snow-clad mountains and a myriad of pools and lakes.

This evening watch was proving truly wonderful, geese appearing from every direction as they embarked upon their regular evening roosting journeys to seek out a haven to spend the night. This is often a mudflat, but that evening with the tide high, it looked like the fields and other open ground further to the south was their ultimate destination.

The rhythmic calls of the geese were captivating: the sound of the sub-Arctic. Watching these dusk and dawn flights of geese on the move to and from their roosts is one of nature's special experiences. It is a cacophony of noise that is part of winter's being, so spellbinding that one could never tire of listening to their calls sweeping across the wildness of the estuary.

Greylag Geese

Over on the far bank a smaller skein of around a dozen greylags crested the near horizon. I uttered a gentle prayer, for about an hour earlier when I was exploring that side of the river, I had encountered three hunters with shotguns, accompanied by dogs, making their way to some secluded spot from where they could ambush the incoming geese. "Please don't fly over the shooters", I implored silently as I watched the greylags wheel in – "please keep safe".

The geese honked in a frenzied manner, and I held my breath as they swooped over. It was like watching a slow-motion drama as I waited for the searing crack of cordite to break the air, and the heart-breaking sight of a goose tumbling out of the of the sky on flailing wings, its body riddled with shot. But no gunfire broke and the geese headed out across the Forth to safety, their calls fading into the distance.

Shortly afterwards and away on the far bank, a torch beam flickered in the encroaching darkness as the shooters made their way back to their car. As luck would have it, and chance was all it was, they had positioned themselves out of the flight path of the geese. Some other evening, it would have been a different story.

As their torch beam glimmered, my mind wrestled in an inner turmoil about shooting, leaving me confused and frustrated in equal measure at my inability to develop a clear and strong viewpoint on the issue. My hypocrisy shone out because I am an angler and a meat eater, but at the same time I have great unease at wild creatures being shot. How easy it must be to be a strident environmentalist that is totally against shooting, as opposed to an indecisive one like me. I have always been a pragmatist, respecting the rights of others to do what they feel is right. Besides, from my experience, people who engage in field sports are almost universally passionate conservationists and knowledgeable about wildlife, and their activities can protect wild habitats.

My somewhat blurred position is one of accepting shooting if it is bagging a bird for the pot or is carried out as part of a carefully controlled culling programme for the greater good of the environment, for example with red deer in the Highlands. If it is for taking birds' lives on a large scale purely for the dubious pleasure of what could effectively be likened to target practice, then I find that harder to get my head around.

However, few things in life are black and white, and while these hunters were hoping to shoot a goose or two for the table, their activities seemed entirely inappropriate in this environment. Here, nature was looking for a place to bed down for the night. These geese had come all the way from their breeding grounds in Iceland or Scandinavia, or perhaps even as far away as Greenland, regarding our shores as a safe place to escape from the cruel northern winter; they should be welcomed and protected.

After all, we have a very special responsibility to look after our geese, given that globally Scotland is hugely important for northern geese populations. Scotland holds around half of the world's wintering pink-footed and Greenland greater white-fronted geese populations, and close to a quarter of its barnacle geese. Virtually all Icelandic greylag geese winter here. This underlines why conservation needs to be international and all-encompassing, carried out on an integrated basis with other countries so that breeding habitats, along with migratory stopover points and wintering grounds are properly protected.

In winter, the Devon estuary comes into its own, a haven for teal, mallards and goosanders and always with the capacity to spring a surprise. The estuary here is unusual, for it is an estuary within an estuary, with the river discharging into the narrow confines of the inner Forth only a few miles downstream from Stirling. The thick banks of mud are one of the most striking features, which in some places sweep up the banks at low tide like a formidable wall. While these mudbanks look forbidding and dangerous, they also form a crucial element of the estuary's lifeblood, for they are home to invertebrates such as amphipods, worms and tiny molluscs.

The other distinctive feature of the estuary is the weir on its upper end, which completely divides the brackish water beneath it from the freshwater above. This weir is the first obstacle that salmon, sea trout, eels and lampreys must negotiate on their way to their upstream spawning grounds. There is a fish ladder in the centre of the weir to aid passage, but from my experience of watching these stepped passes, salmon and trout only use them by chance and are as likely to use any other part of the falls. When the tide is high, fish have a relatively small rise to breach, whereas at low tide it is nigh-on impossible to make successful passage such is the size of the drop.

Interestingly, I once encountered an angler who on several occasions had caught flounders on baited hooks on the middle river, several miles upstream from the estuary. How a flatfish is able to cross the weir is completely beyond me, although it does exhibit the sheer resilience and tenacity of some of our wild creatures. It is well known that flounders, which are essentially sea fish, often enter rivers, but I've never fathomed the reason why, especially since the feeding out at sea would be much richer.

One consequence of the weir is that salmon and sea trout congregate in the pools beneath it waiting for the tide to rise high enough to make good their passage. This offers opportunities for predators, and there is one that outclasses every other when it comes to strength and power – the grey seal. They are not regular visitors to the Devon estuary but are most likely to be seen in September and October as they intercept incoming salmon.

Grey Seals

Just a couple of months previously, I had encountered a seal at high tide making its way slowly upstream by the east bank. It was a bull seal, powerfully built with a heavy head and long snout. Each time he surfaced, an audible exhalation of air drifted across the breeze, before he rolled and dived under once more on the hunt for salmon.

Grey seals are formidable predators, and very adaptable, too, being equally at ease at fishing many miles offshore in the North Sea, as in the shallows of the estuary. Seals have an inner depth that is absent from many other creatures. I recall one summer long ago rowing a small dinghy across a west coast sea loch with a grey seal following purposefully in my wake. What was it thinking? Was it apprehensive, or was it relaxed by my near presence? It was hard to tell from the seal's inscrutable expression, but within those liquid eyes there was a penetrating intelligence that shone out at me. My inescapable conclusion was that the seal was following me out of sheer inquisitiveness.

The most unusual fish predators I have ever come across by the Devon estuary was a pair of porpoises, which I sighted several years previously just off the mouth of the river. Out on the Forth, I noticed a submerged bow wave rolling through the water, a bit like a torpedo trail. My initial thought was it was a seal, but then a triangular fin of a porpoise broke through the muddy water. My goodness, this was incredibly far up the

Forth for a porpoise to occur, what on earth was it doing?

It was hard to be sure, but it looked like all was not well with this little cetacean, for it kept on surging towards the shore, sometimes grounding in the mud, before clearing itself with an energetic flap of the tail. Shortly afterwards, another porpoise appeared in the middle of the channel, almost as if keeping a watchful eye on its companion. Could this be a mother and a well-grown calf and one of the pair was poorly? Sadly, this indeed was the case, with the strangely behaving porpoise succumbing a few days later.

Harbour Porpoises

Porpoises are undeniably attractive creatures, so refined and well adapted to their watery environment. It is hardly surprising that we have such a special affinity with these animals. I'm especially fond of their endearing nickname 'puffing pig', which refers to the sneeze-like sound made when breathing.

For me, one of the most compelling aspects of the Devon estuary is the incongruity between the signs of humankind all around, contrasting with the wilderness of the river – nature and people co-existing in relative harmony. Whisky warehouses encroach upon the northern and western edges of the estuary, and on the south bank of the Forth lies the old, and long decommissioned, naval munitions depot at Bandeath. This munitions area, reflecting times of past conflict, consists of an accumulation of eerie grey buildings, each one sited a reasonable distance apart, presumably in case one of the storage facilities should have ever exploded, thus reducing the chances of a catastrophic chain reaction. Looking downstream along the Forth, the fringes of Alloa loom into view, making this very much a place where the imprint of humanity is well engrained.

Despite this, the Devon estuary and associated inner Forth have a wonderful natural aura, a place of ducks bobbing out on the water, cormorants diving in search of fish and gulls wheeling in the air. The nearby Cambus Pools on the western bank is a Scottish Wildlife Trust reserve, a rich reedbed that holds sedge warblers, reed buntings, moorhens and water rails. By the other bank, there are other more ephemeral large ponds, known as the Devonmouth Pools, creating a rich oasis for wildlife and a place where shelduck often linger.

The true stars of the Devon estuary in winter are the teal, such attractive little ducks, so perfect in form and their flute-like whistling calls completely addictive to the ear. The early twentieth-century field naturalist Thomas Coward in his *Birds of the British Isles and their Eggs* (Frederick Warne & Co. Ltd, 1950) described their vocalisations so eloquently:

There are few more talkative ducks than the teal; birds in winter flocks chuckle conversationally, and on the meres, the loud clear call, a short sweet whistle, rings out incessantly. When the drakes are courting the low double whistles run into a musical jumble, a delightful chorus.

As December turned into January, I regularly ventured down to the estuary, keen to achieve a close sight of the wildlife, and each time it was the teal that enraptured me by their busy activities, often feeding by the muddy water's edge, their bills eagerly sifting the glutinous ooze for minute creatures.

The estuary is a sanctuary for waterfowl because it remains ice-free no matter how cold the weather turns, and food is forever accessible.

Teal are shy and always on the alert, and on one visit, I watched a peregrine falcon swoop above the estuary, scattering the ducks into the air. Teal are a favoured prey for peregrines, but they are challenging to catch when in flight, as they keep together, turning and twisting all the time, and engaging in spectacular corkscrewing dives.

On a mild day in early January, boisterous groups of teal engaged in

Peregrine Falcon

what I presumed was early season courtship, the drakes and females chasing each other in short flights, before splashing back down into the water again. Teal are mainly winter visitors here and once spring takes hold, they move off to their breeding grounds in bogs, pools and reed-fringed lochs throughout Scotland and beyond.

Another regular winter visiting duck on the estuary, although in much smaller numbers than teal, is the goldeneye. A winter migrant from Scandinavia, goldeneyes are nervous ducks that often take to the air on whistling wings at the first hint of danger. One afternoon, as I brought a drake into focus through my binoculars, I was immediately struck by his bright yellow eyes which caught the sunlight in a most remarkable way. It was almost as if the eyes were powered by electricity, their glowing orbs shining out like beacons. These ducks certainly have a most appropriate name and those gleaming eyes have a burning intensity that is most compelling.

Estuaries by their very nature are unique and special places, nourished by a continual two-way flow of nutrients from both sea and river that makes a typical estuary many more times productive than a farmer's cereal field. The productivity figures are quite astonishing. In the Ythan Estuary, north of Aberdeen, research has

Goldeneye

shown a square metre of mud is home to tens of thousands of minute shrimps, mud snails, ragworms and tiny clams. There can be few, if any, other environments in the world that can match such richness.

From my winter wanderings, I concluded that the Devon estuary was most likely on the lower end of the productivity scale, probably because it does not spill directly out into the sea, and the water has lower salinity than many other estuaries, making it less suitable for marine worms and other mud-living invertebrates to thrive. This is borne out by the paucity of wading birds such as redshanks and curlews on the estuary. A few miles further down river, and towards the sea, at places such as Culross and Skinflats, much greater numbers of waders occur.

Ragworm

Nonetheless, I could not be absolutely sure about the abundance of invertebrates, and one January morning after an enjoyable hour watching herons prowling the banksides, teal dabbling in the shallows and a determined goosander successfully grappling a flapping flounder, it was clear that the only way to find the answer was to get up close and personal with the estuarine mud and embark upon some sampling. It was an unappealing thought, for the mud is thick and heavy, with an unpleasant algal aroma.

Redshanks

Furthermore, I had concerns about my wader-clad legs getting trapped and sucked deep into its dangerous clawing embrace.

I was also keen to find out more about the other underwater creatures of the estuary – and that meant snorkelling. Again, this was an uninviting proposition, given the extreme poor visibility of the silt-laden water, as well as the potential trapping hazard presented by the mud. Such exploration bordered on the foolhardy, and given the cold, January was certainly not the time to take the plunge. I would need to plan carefully such a venture, identifying the safest route in and out of the water.

Thoughts swirled across my mind. I instinctively knew it had to be done, and the right time would be in the spring when the water was warmer and more clement. The decision to delve deeper into the estuary sparked a frisson of excitement through my body, thrilled at the prospect of unveiling more of the river's hidden secrets over the coming months.

Chapter 14
ON THE TRAIL OF THE ELUSIVE OTTER
(January)

Since my initial brief encounter with a pair of otters at dawn back in December, I struck lucky once more on a section of the lower river the following month when out checking my trail cameras one afternoon. There was a disturbance in the water by a fallen alder on the far bank, so I stopped to investigate, unsure what type of creature had caused the commotion.

In the tangled branches a few feet above the water's surface, the rhythmic white flash from the nervously twitching tail of a moorhen caught my eye. The bird was worried: something had spooked it. Then, a surge of water

Otter

and up bobbed the head of an otter, so close that I could discern the glint in its eyes and its wet whiskered face.

It immediately spotted me, and with discernible astonishment in its eyes at the unexpected sight, dived under again in a large swirl of agitated water. It was surprising behaviour from an otter because normally they are more relaxed when they spot a person, confident the water provides a safe sanctuary. I think in this instance, the animal had seen me suddenly and in proximity, which sparked its desperate dive under the water.

I waited. Bubbles rippled the surface. "Please reappear", I prayed. "Please show yourself"; for daytime sightings of river otters are rare and so very precious. The stream of bubbles continued to ruffle the water, but time ticked on and no otter materialised.

A sigh of disappointment: the animal had made good its underwater getaway, and rather than the bubbles being produced by the otter, I concluded they were a mere coincidence, probably caused by gas being released by decomposing vegetation on the riverbed. Indeed, I had witnessed at first hand such riverbed bubbling several months before when I had snorkelled in this languid stretch of the river.

Despite the fleeting glimpse of the otter, the encounter was enthralling, the memory of its softly curved face lingering long into my consciousness. My mind was also full of regrets in that my approach along the bank had not been more cautious. If only I had looked further ahead as I went. But observing nature is full of 'if only' moments and it is very much a three-dimensional process, with your eyes simultaneously scanning the skies for birds, and the surrounding vegetation and the ground for plants and animal signs.

My trail cameras were picking up otters on a regular basis, including a pair lolloping about on a sandbank during a snow shower in January. It almost seemed as if the otters were curious and enjoying the snow, such was their boisterous behaviour.

This pair could have been a mature male and female, although for the most part adult otters are solitary, and will only frequent each other's company for a few days prior to mating. Most likely, this was a mother and a nearly full-grown cub, and they seemed at great ease in each other's company. Young otters can stay with their mothers for up to year, providing a valuable opportunity for them to watch and learn the most effective hunting techniques.

Unusually, otters are one of the few river creatures that can give birth in any month of the year, although most in Scotland have their cubs in the summer.

Although not exclusively so, otters on the Devon are mainly nocturnal and it is always a source of mystery to me how they can catch fish in the dark, or in the murky waters of the estuary. They do so through the aid of their long whiskers, known as vibrissae, which are sensitive to vibration and literally feel for the fish in the dark void. While fish form the bulk of their diet, otters are eclectic in their tastes, and will also consume frogs, small mammals, eggs, and nestlings, as well as aquatic invertebrates.

The mainly nocturnal habits of the Devon otters contrast starkly with those living on the west coast and the Northern Isles, which are often active by day. I presume this is because such areas are quieter and less disturbed, and their feeding habits are likely to be influenced by the tides.

Only the year before, I had just disembarked from the ferry at Lerwick in Shetland, to immediately spot a large male otter in the water below the quayside. It was holding a small shore crab in its front paws, which it swiftly despatched between its jaws with a loud crunch, leaving behind a scattering of shell fragments that spiralled gently down through the clear sea.

My initial hope on the Devon was to find an otter holt (den) with cubs, which would provide the perfect opportunity to view their behaviour. After my wanderings along the river of the previous few months, reality eventually took hold that this was not going to be possible. The chances of finding a holt were minimal, given the plethora of tangled alder roots along the banksides. Every hole, crevice and cavity had the potential to be a holt, and there were literally thousands of them.

No, the only reliable way to detect the presence of otters was to search for their spraints. Otters mark their range with spraints, depositing them on prominent places, such as on rocks. They are used as a means of communicating with other otters, indicating their presence, and used as signals of their breeding condition.

I laid down plans to watch otters in the coming spring, picking sites on the river that offered protracted down and upstream views during the longer daylight hours, which would greatly enhance my chances of observing one at dawn and dusk. However, as things turned out, that ultimately proved impossible due to the 'Coronavirus Lockdown' restrictions that were imposed at the end of March.

Fortunately, I did chance upon a wonderful view of an otter one dawn morning at the beginning of June as my river year reached its conclusion. As I made my way along the riverbank, I noticed an otter swirling in the water on the far bank around a small willow thicket. A willow had tumbled into the water a year or so back, and from its submerged trunk, new sapling shoots had miraculously sprung up above the surface, creating a mini island by the bank. The otter tumbled and weaved around this leafy islet, diving under, its back rolling in an arc, and the tail sometimes pointing straight out of the water.

This was the exact same behaviour I had twice previously seen otters engage in, and is evidently an essential element of their foraging, disturbing sediment from the riverbed and perhaps attracting small fish such as minnows in the process. I wonder whether invertebrates that had been stirred were also being snapped up.

Occasionally, this otter paused in its swirling and tumbling, and looked across curiously at me with intelligent eyes and its long whiskers glistening in the dawn light. Soon it tired of its foraging, and with one last glance at me, dived under. I looked up and down both sides of the bank, but it never reappeared, having merged seamlessly within the bounds of the river.

Chapter 15

CATKINS, HERONS AND ELECTRICITY
(January–February)

The lime-tinted glimmer of catkins adorned the bare branches of this hazel beside the upper river by Rumbling Bridge in Kinross-shire. By my feet, more green verdant freshness: the emerging spears of snowdrops heralding the approach of a new season.

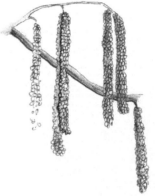

I heaved a sigh of contentment, for the riverbank was now on the cusp of remarkable change – a time of renewal and anticipation. It may have been mid-January, but winter's grip was already yielding, albeit reluctantly. Frosts, snow and bitter winds might still lie ahead, but in an imposing battle of wills, the lengthening days and stronger sun would eventually gain the upper hand.

I examined the catkins and brushed my fingers across a cluster. They were hard and smooth to the touch, but over the coming

Hazel Catkins

weeks would lengthen and become fluffy, by which time they are most appropriately known as lambs' tails. These dangly green catkins are the male flowers, but as I looked closer at the branches, the tiny emerging bud-like red female flowers could just about be discerned, too.

Pollinated by the wind, these develop by autumn into small hazelnut clusters; although productivity never seems to be great and most hazels in autumn seem to be completely devoid of nuts. Those trees that do bear nuts are a magnet for agile wood mice and bank voles, the neatly chiselled shells lying below being testament to their nocturnal foraging activities.

Wood Mouse

I wandered further upstream along a stretch of river a few hundred yards above the spectacular rocky gorge at Rumbling Bridge. Alders were also beginning to produce catkins, although they lacked the vibrancy of those found on hazel. However, alders bring colour in a more extraordinary way, for when a tree cracks and tumbles, the inner wood displays a stunning bright reddish-orange hue. At one time this striking feature of new-cut wood was thought to symbolise bleeding and as such the tree was shrouded in fearful superstition. Over time, this deep colour fades to a paler brown, making the wood much sought after by furniture makers who, quick on the marketing uptake, brand it as 'Scots mahogany'.

The alder is our quintessential riverside tree and copes well in such poor and waterlogged conditions, and through symbiotic relationship with a type of bacterium has the capability to absorb nitrogen from the air, which in turn enhances the surrounding soil fertility. It is a pioneering tree that colonises areas of boggy ground, which over time create the right conditions for the succession of other types of vegetation.

Alders help to prevent erosion of riverbanks, their spreading and probing roots binding the soil together, and, like the hazel, is a tree that should be revered as one of our shining stars of nature. In winter, feeding flocks of siskins and redpolls dance along the riverside alder tops in search of their small seed cones. These little finches twitter in joyful harmony and no matter how ferocious the weather, the alders are always productive and willingly give up their precious bounty.

Shortly before I reached Crook of Devon – a village where the river suddenly changes its south-easterly course and starts to head westwards – I disturbed a heron from the bankside, which took to the air on slow lumbering wings, uttering a harsh 'kerr-ack'.

Herons are typically shy birds, so I was not surprised that this one had taken flight so easily. Not all behave this way, and one I had come upon a few weeks previously had no intention of flying away, despite my close approach, thus providing an opportunity to study it in detail. It was indeed an impressive bird with its yellow dagger-like bill and pendant crest combined with dark markings down the front and neck.

Herons have a distinct air of the primeval and watching one take flight on a mist-laden morning down by the river imprints upon the mind surreal imagery of a pterodactyl disappearing into the grey wispy murk. They are

Grey Heron

wily and will often use village streetlights as an aid when hunting fish at night along the tributary burns of the Devon. I suspect they similarly utilise the soft white glow of the full moon for nocturnal forays in areas away from our towns and villages.

They also quickly learn the best fishing spots, often at the top of a riffle in a river where they can snap up tiring trout as they move upstream. They are adaptable too, and in spring, frogs form an important part of their diet. It is thought that egg laying is timed to coincide with the abundance of spawning amphibians, providing a bounty of food at a crucial time. Herons will also take mice, voles and large insects such as dragonflies.

Dragonfly

Much folklore promotes the notion that herons have some special attraction to fish. It was said, for example, that the marrow from the thigh bone of a heron when applied to a baited hook would help anglers boost their catches. Others believed that certain oils extracted from dead herons were similarly effective in luring fish. The truth is that rather than relying upon inherent bodily attractants, it is stealth and guile which makes the heron such an efficient fish catcher.

The most engaging tale, however, is the delightful image conjured by the long-held belief that the floor of a heron's nest has two strategically placed holes to enable the long legs to dangle through when it is incubating eggs.

Herons always have the capacity to surprise, and on one January afternoon in a stubble field adjacent to the lower river, I stumbled upon a gathering of 11 birds resting up. This was a communal roost, and a sight I had only previously witnessed on a handful of occasions. There was no interaction at all between the herons as they gently slumbered, but I imagine they gained comfort by being together with their own kind.

A few weeks later in February, I returned to Rumbling Bridge and spent some more time exploring the gorge. It is a truly impressive place and a most unusual geological feature: a deep narrow chasm where the Devon thunders and roars before releasing itself upon a much calmer stretch below the village of Rumbling Bridge. From there, about a mile or so further downstream, the river then cascades over the precipitous Cauldron Linn.

The deep rocky cleft at Rumbling Bridge is rich in moisture and shade-loving mosses and liverworts, perpetually showered with fine spray-droplets from the raging water. River flow is a powerful thing, and a few hundred yards upstream from the gorge, there is a community hydro-electric scheme for local power generation. The construction was completed in 2016, and if the water flows at a reasonable rate, the scheme generates a steady 500kW of electricity – just one more example of how rivers benefit society.

All this got me thinking about the fast-flowing tributary burns of the Devon much further downstream, which tumble through the deeply incised glens of the Ochils at Alva, Dollar, Menstrie and Tillicoultry. In the eighteenth and nineteenth centuries, these gushing burns were like gold dust to these villages, for they played a key role in powering the Industrial Revolution in the form of textile production. For the wild creatures and plants of the Devon, it was to prove a time of great turmoil because the dawn of industry meant pollution and a whole new raft of challenges; the era of clear and pure flowing water was coming to an end.

Chapter 16
THE RIVER'S GREATEST CHALLENGE
(February)

Early February, and as the gentle breeze ruffled through the treetops of this small wooded ridge above the Devon near Tillicoultry, I breathed in the wildness of this quiet and serene place.

All was not as it might seem, however, for when I stooped down and clawed the soil, tiny fragments of coal clung to my fingers. I was standing on a spoil heap from a redundant coal mine, yet nature had largely consumed these past industrial workings, leaving behind only traces of how the landscape once looked.

Wild Strawberries

Birch, sycamore, ash, alder and willow all abounded. Wild strawberries thrived here and there were other surprises too, including broad-leaved helleborine, a tall and rather striking orchid which flowers in late summer.

I find this inherent ability of nature to reclaim past losses so reassuring. Coal was mined in this area extensively until the mid-1960s but today it is wildness that reigns supreme. Not so long ago this would have been a noisy and bustling place, with coal wagons trundling to and fro and the excited shouts of miners and the constant clings and clangs from the excavations below ground. Such clamour had been replaced by the gentle winter warble of a nearby robin.

Broad-leaved Helleborine

It was a telling reminder of the past and how the local landscape had undergone fundamental change over the last few centuries, with pollution from woollen mills, bleachfields and the wash-out from mines causing havoc upon the river's wildlife.

The burns that tumbled down the southern scarp of the western Ochils had always been an essential source of water for communities developing along the base of the hills. By the late seventeenth century, they also became an increasingly important driver of power with the first corn and grain mills appearing, such as at Alva.

From as far back as the sixteenth century, the weaving of plaids and course blankets was a cottage industry, with women washing the wool and the men weaving by hand. At this time, the area was on the cusp of the Industrial Revolution, and from the late eighteenth century onwards, many water-powered woollen mills sprung up at Menstrie, Alva and Tillicoultry. Textile production became the core element of the local economy, producing serge, blankets and shawls.

The fast-flowing hill burns were channelled to drive water wheels, and dams and lades were built to ensure there was enough water to last all year. It was a period of dramatic change, and Alva and Tillicoultry developed so rapidly that their populations increased faster than the infrastructure could support. In 1755, the population of Tillicoultry was 787, and by 1881, it had soared to 5,544. However, the burns were always limited in flow, and the abundance of local coal led to the eventual transformation to steam power.

By 1850, there were over 40 woollen mills in these Ochil towns, popularly known as the Hillfoots, but afterwards a gradual decline ensued, and by the second half of the twentieth century only a handful remained.

The impact on the Devon of such rapid industrialisation was significant. William Gibson in his book *Reminiscences of Dollar & Tillicoultry* (Strong Oak Press, 1882), recounted polluted water entering the Dollar Burn (a tributary of the Devon):

Behind the second wool mill (which is the only one I remember of) there was a large, deep pond, for running the dirty Waulk Mill water into; and this was emptied occasionally during the night, and the burn not polluted when people were requiring water...The inhabitants of Dollar were, I believe, indebted to the

Rev. Dr. Mylne for the construction of this pond, and preserving to them their beautiful stream of pure water. Previous to him coming to Dollar, the dirty Waulk Mill had been regularly run into the burn and greatly polluted it; and against this nuisance the doctor sent a strongly-worded protest to the manager of the mill and insisted it be stopped at once.

It is just one example of pollution entering the river system, and such instances would have been replicated on a much larger scale from the more heavily industrialised villages further downstream. Coal mining also contributed to the flow of pollutants and the resultant increase in population from such industrial expansion would have led to vastly increased quantities of human waste entering the river. Sewage pollution can cause rapid algae growth, starving rivers of the oxygen that insects, fish and other wildlife need so desperately to thrive and hitting other species further up the food chain.

It is likely, too, that the rising population brought additional pressures in the form of poaching. In the *Statistical Account of Scotland* (1791 to 1799), the entry for Dollar noted:

About 20 to 30 years ago, salmon were found in Dovan [the Devon] in great plenty; but, from the illegal murderous manner of killing them with spears, their numbers of late have greatly decreased.

The *Account* also stated that:

The river, being small, does not admit many kinds of fish in it; yet there are very fine freshwater trouts, of a considerable size taken in it, as well as parrs, in great numbers. In harvest, sea trout are likewise killed in it, from 2lb to 4lb weight. And, in the season, salmon are caught from 5lb to 20lb.

This assertion suggests that for much of the eighteenth century, the Devon was still relatively clean and holding reasonable populations of fish, but pressures were mounting with each passing year. Indeed, in the second half of eighteenth century, the engineer James Watt proposed to canalise the Devon over a four mile stretch from Dollar to Sauchie for the purpose of transporting coal and other minerals being mined in the area at the time. The scheme never saw the light of day, but had it gone ahead, then the river would have undergone fundamental and seismic change.

From the start of the nineteenth century onwards, pollution from the rapid growth in textile production, mining and other industries began to seriously impact upon the natural diversity of the middle and lower river. It was a familiar problem replicated in several other parts of Scotland. Nigel Holmes and Paul Raven in their book *Rivers* (British Wildlife Publishing, 2014), recounted how well into the 1900s, sewage and industrial waste from Selkirk's ten mills in the Borders went down in the drain and straight into the River Ettrick, a tributary of the Tweed. Downstream, the river often ran bright blue (from dye) and in 1906 was described as having a 'deep black colour'.

While the wildlife of the Devon suffered hugely, it probably never became a truly dead watercourse because the upper sections of the feeder burns, as well as the higher river itself, ran pure and clear, and during times of spate, the river had the ability to flush and temporarily cleanse itself.

Indeed, William Gibson in his 1882 book of reminiscences remarked upon the abundance of trout in the hill burns of the Ochils, including the Daiglen Burn and the Gannel Burn above Tillicoultry.

Moving to more modern times, the minutes from the Devon Angling Association (DAA) in 1926 reveal concern among anglers about the extent of pollution on the river, including sewage at Alva, coal dust from the Devon Colliery and dye entering the burn at Tillicoultry. Minutes from 1941 stated that pollution on the river was "still bad from Tillicoultry downwards". In 1946, the DAA reported that hundreds of dead fish were found in the "the stretch of river from where the Dollar Town Council Sewage effluent ditch enters the river down to Rack Mill Bridge". In 1949, the DAA claimed that the "Devon was polluted at low water and certain parts showed very little fly life. Up to Dollar fly life was below what it ought to be principally as a result of domestic sewage". Minutes from 1956 revealed:

> *The Secretary [of the DAA] reported that he had written to the Forth River Purification Board asking whether the disgracefully polluted state of the lower reaches of the Devon was likely to be improved upon and that he had received a very exhaustive report on the position which indicated there would be a lapse of five years before pollution was completely cleared but that action was already*

*being taken at Cambus and that an improvement could be expected in about a
year's time. The board admitted that the pollution was particularly bad from
Alva downwards and the Glenochil Distillery [a yeast factory] waste was
particularly deadly to fish life.*

The build-up of organochlorine pesticides from agriculture also became
a widespread problem in many rivers in the 1950s, affecting animals at the
top of the food chain such as otters, whose numbers plummeted during
this period. In the two centuries preceding, it is also likely that otters were
persecuted on the Devon because of their perceived threat to valuable
salmon and trout, and possibly also for their pelts.

Keen to get a personal perspective of the river during these more recent
times, I caught up with angling enthusiast, Bryan Anderson, to find out
more. Born in Sauchie in 1944, he started fishing the Devon in 1956 and
recalled various industries operating close to the river. He told me:

*One of the main polluters was the paper mill at Tillicoultry which operated
between 1921 and closed in 1972. Of the mines, the worst one was the Devon
Colliery near Fishcross: the reason being they had a settling pond adjacent to the
river which leaked coal dust constantly into the river.*

*The paper mill leaked coloured dye into the river: if they were using purple
dye, the river turned purple, and if yellow, the river turned yellow. This
continued on a regular basis.*

*Despite this, we still managed to catch fish in these polluted times. As a boy I
lived in Sauchie and my buddies and I would go down to the Black Devon
[another nearby river] to trap minnows which we used as bait on our weekend
trips to the River Devon. We would fish the diving minnow during daylight,
then at night just let the minnow settle on the bottom. We had trout up to 2lbs 8
ounces by this method and we also had a back-up can of worms and grubs if the
minnow bait was not catching anything.*

*Even though the river was polluted we always caught trout in the two main
bits we fished.*

He also remembered there being a great abundance of eels in the river
at that time, with one deep pool being so full of them, that every cast
would result in one being hooked. Eels have a greater tolerance to pollution

compared with other fish, and perhaps their abundance reflected the state of the river then.

Other fond memories expounded by Bryan included catching a 12lb salmon – the biggest he has ever caught in the Devon – and in the early 1970s watching 'a swarm of lampreys' brush past his waders. He has never seen the same abundance of lampreys since.

River quality across Scotland has improved markedly in recent decades, most notably since the agreement of the European Water Framework Directive in 2000, which was enabled into Scottish Law in 2003. Furthermore, targets have been set to ensure most of our rivers are classed as having reached a good or high ecological status by 2027. Time will tell whether such a target will be achieved, but ambition is essential, for rivers are one of the bedrocks of our precious environment.

While the Devon today is largely clean and healthy, it is still not without its pollution incidents, most particularly from occasional overflows from sewage treatment plants, which is legally permitted on a limited and regulated basis, especially after heavy rainfall so as to prevent back-up flooding, as long as it is deemed that such discharges do not ecologically affect the river. Despite this, I find it astonishing that in this modern era such discharges are still permitted: it seems to be a nineteenth-century solution to a twenty-first-century problem.

A further problem today on a few parts of the Devon is the seepage of iron oxide from redundant coal mines, causing an unsightly orange flow. The run-off from agriculture of nutrients and pesticides, as well as drainage from roads, are other perennial issues currently facing our rivers, including the Devon. And then, there is plastic. After the river recedes following a spate, it is depressing to see the amount of plastic debris littering the banksides.

Weirs also negatively impact upon the river. A legacy of our industrial past and built to hold and control the use of river water, most often for mills, they now create formidable barriers for the passage of trout, salmon and other creatures. Fish ladders work to a degree, but trout, salmon, eels and other fish in all our rivers would receive a huge boost if there was a concerted strategy to remove weirs wherever possible.

It is the cumulative effect of all such issues that is the real challenge facing our rivers today – each one adding up to create a greater whole.

The story of the Devon is of a river that has largely recovered after years of neglect; a microcosm of the inherent ability of nature to bounce back if given half the chance to do so. But it is not a pure and perfect river, and sadly never will be, because it still suffers the scars from the past, and the daily impact of humankind will forever haunt it into the future.

Chapter 17
IN SEARCH OF RATTY
(February)

'It's my world and I don't want any other.
What it hasn't got isn't worth having and what it doesn't know is
not worth knowing. Lord! the times we've had together!
Whether in winter or summer, spring or autumn,
it's always got its fun and excitements.'

Kenneth Grahame, *The Wind in the Willows*

Sadly, the real-life story for Ratty, the engaging water vole so evocatively portrayed in *The Wind in the Willows* by Kenneth Grahame, is not one of riparian rhapsody but rather a sorry tale of tumultuous population decline that has seen numbers plummet catastrophically over the last few decades.

Water Vole

Whilst riverside habitat degradation has been an important factor in many areas, the principal cause for this dramatic fall is predation by the American mink. When danger threatens, a water vole dives underwater or seeks refuge in its burrow, but the mink is an accomplished swimmer and can pursue its quarry with great ability.

I've read in wildlife literature that female mink are slim enough to even pursue the hapless voles down into their burrows. I find this unlikely, as I do not think they are that petite; however, a mink would easily be able to dig into and enlarge vole tunnels, to gain access.

When I first embarked upon my year-long river journey, I was more than aware that water voles were absent on the main course of the Devon, for I had never seen one in all my years wandering the banksides. While mink predation was the underlying cause, I also presumed that these delightful creatures were never previously abundant here, given that the Devon is a spate river, and the voles' bankside burrows were liable to be flooded out. Certainly, the creatures seem to favour slower-flowing water courses, neither too wide nor deep, which leads them to prefer backwaters, oxbow lakes and burns, rather than large rivers.

Presumptions, however, are often incorrect, and I have since learned from several conversations with people who lived in the area at the time, that water voles were indeed present on the middle and lower reaches of the Devon up until around the mid-1980s.

As a youngster growing up in Edinburgh in the 1970s, I recall vividly water voles being abundant on a lower stretch of the Water of Leith at Warriston. I adored watching them from my vantage point on a disused railway bridge that looked directly down upon the water.

These Edinburgh animals were cuteness personified: plump and rotund, with chubby cheeks – a bit like miniature beavers. Most mammals are shy and nocturnal, but not these brazen water voles, for they went about their business in broad daylight, right next to a busy road. Their burrows pock-marked the far bank of the river, and it was common to hear their distinctive 'plop' as they dived into the water, followed by their V-shaped wakes rippling the surface, little legs whirring away like miniature paddles.

Then, the mink arrived, and the water voles soon disappeared – a story that was replicated on rivers across Scotland, including the Devon.

While now gone from the main river, several years ago I spotted a water vole swimming by the edge of a large pond about half a mile away from the Devon, and a friend of mine had encountered one in his own garden pond, which again was relatively close to the river. Could water voles still be hanging on in the ponds, burns and ditches adjacent to the Devon?

In mid-February, I searched for their signs along two likely-looking drainage ditches that fed into the middle river. One of the best ways of finding their presence is to seek out their feeding stations, where the remains of chewed vegetation, such as rushes, accumulate. Their droppings are also distinctive, which are deposited in latrines. Alas, I could not find any signs, but part of that could have been down to the torrential rain that fell for much of February, hiding their signs under the unusually high-water levels. Undeterred, I set three trail cameras in different parts of the furthest reaches of the drainage channels, as far away from the river as possible, where mink predation was least likely and where the water level was not too high.

I retrieved the cameras several days later. As I feared, there were no water vole captures, although one of my cameras had picked up a water shrew swimming along the edge of the ditch; and one filmed a brown rat crossing the water.

Water Shrew

Water shrews are seldom seen creatures that are avid predators of a wide range of small creatures including, invertebrates, minnows and tadpoles. They will even take frogs, which is an amazing feat, given that in comparative terms, it is like the titanic struggle involved when a lion tackles a buffalo. Mild poison in their saliva is thought to help water shrews subdue their prey, and they are capable divers, foraging for prey along the bed of a burn, ditch or pond. Well-designed for their aquatic lifestyle, the haired tail acts like a keel or a rudder, and the bristles on the feet aid paddling.

Slightly larger than a common shrew, water shrews have a most distinctive two-toned pelage, almost black on the upper side, contrasting sharply with its white underbelly. Several years ago, I came upon a water shrew by the edge of a tributary burn in Glendevon that was in its final death throes. It was unsteady on its legs and staggered around in circles, continually falling over onto its side. Shrews are short-lived animals, but I was nonetheless saddened by the undignified end for this fascinating creature.

I also recall, many moons ago, watching a water shrew by a burn in woodland on the southern edge of Loch Earn in Perthshire. It scampered along the edge of an exposed cut in the bank, before disappearing down a tiny burrow. I waited for a while, and it emerged again and slipped into the water where it dived under, the body turning silvery from the trapped bubbles in its fur. Then, it disappeared, presumably surfacing under a mossy overhang where I had no clear view.

As for the brown rat, well, they are one of those animals that are much maligned and universally detested. Some of this perception of them being dirty and disease-carrying animals is unfair, for while they are common inhabitants of sewers, rubbish tips and buildings, many also live a cleaner life in the countryside, and are frequent dwellers along watercourses where they are adept swimmers. Water voles are often called water rats (hence where the name 'Ratty' comes from), but they are not rats at all, and are as pure a vole as any other type of vole in the land, albeit larger and with a longer tail.

The brown rat, like the mink, is another one of the many non-native species that lives by the river. Originating in Asia, they first arrived on our shores in the first half of the seventeenth century via ships. It is one of our great urban myths that you are never more than several feet away from a rat at any one time, which if true would result in our stumbling across rats all the time. Whilst their abundance may be an exaggeration there is no doubt that rats are highly successful animals and equally at home in town or country.

I do wonder whether rats are potential competitors with water voles, given that they often live in the same areas. And, why were rats able to live in this riverine drainage ditch and survive mink predation, whilst water voles could not? Perhaps it is because rats are feistier than water voles. I once saw a cat corner a rat in an urban doorway, only for the rat to turn upon the feline and see it off. I've seen a rat do the same to a weasel.

While the water shrew and the rat were fascinating finds, the lack of water vole evidence on the Devon and its environs confirmed what I had already suspected, and their plight countrywide does look bleak. But there are opportunities to restore numbers. I recall vividly a decade or so ago being invited to visit a water vole reintroduction scheme in Loch Ard Forest in the Trossachs. Here, captive bred voles were being released back into the wild in an initiative managed by the then Forestry Commission Scotland, in partnership with others. The voles originated from animals captured from a construction site near Glasgow, which was close enough to their final Loch Ard Forest release area to ensure a suitable genetic profile.

So that the voles could prosper, the local mink population on the Duchray Water and its tributaries was eradicated. This was feasible in a localised area because the forest lies nestled in a half-bowl rimmed by hills, thus mink trying to recolonise from the one open flank on the upper Forth by Aberfoyle could easily be trapped.

One other key to success was the creation of a diverse network of suitable habitats. Areas along burns where trees have been harvested were kept open and managed to ensure a diversity of flora. New ponds were also created. Such work also benefited a vast array of other wildlife, with the water voles now thriving in the area.

Interestingly, ecologist Rob Strachan in his book *Water Voles* (Whittet Books, 1997), pointed out that mink and water voles can co-exist – provided the environs are right, including rich and lush vegetation comprising a patchwork of aquatic habitats, including reed fen and willow carr. In other words, protect the right aquatic habitats, and even with mink present, water voles will have a chance to thrive.

Water voles are adaptable, and in some parts of Scotland, such as in Glasgow, they do not rely upon water at all, living a largely subterranean existence in dry grassland. They can also live in mountainous areas, and I've come across them when camping by rivers and burns in the high Cairngorms, where the animals are much darker-furred than those found in low-lying areas. Indeed, water vole remains are sometimes found at golden eagle eyries, indicating that the remote upland burns of the Highlands are one of their last strongholds.

The plight of our precious water voles makes me worry for the future. They are such wonderful animals and so charismatic and an example of humankind disrupting the natural way of things. As the months progressed, and the more I got to know the Devon, the more I was becoming downhearted at the way it had become unnaturally altered. Such gloominess, however, was in the most part fleeting, for there were so many positives and wonders of nature lying within the river's bounds – and they outshone the negatives by a country mile.

What would Ratty have made of it all? While he loved the river, he was aware of the dangers: *"Weasels – and stoats – and foxes – and so on… Well, you can't really trust them, and that's the fact"*. He was right, but in the mink, Ratty had met a foe that made all the others look like pussycats.

Chapter 18
THE MIGHT OF THE RIVER'S REACH
(March)

Where does a river begin and where does it end? In a linear sense that is easy to answer – the source and the estuary. But the influence of rivers is much more expansive than that, and one only has to think about the numerous tributaries that flow into the main river, and the even larger number of smaller tributaries that then flow into them, and so on.

While the main river could be likened to an artery, the tributaries and sub-tributaries are the equivalent of smaller blood vessels and capillaries, reaching every part of a body. Accordingly, the extent of a river's reach and its influence on the environment is truly all-encompassing. The river *is* the environment: its beating heart, which drains, nourishes and sculpts landscapes, providing habitat and life. The influence of a river is much wider than the flow contained within its banks, and this is no better illustrated than the wonderful pools and ponds on the haugh in the middle section of the Devon.

It was early March and I knew the frogs would be stirring, preparing to mate and spawn in these pools, which are replenished every time the Devon bursts its banks. I was keen to catch frogs in the act of spawning here, for these amphibians have long been a source of fascination to me.

On arriving upon the flood plain, it was like witnessing the remains of a battlefield: debris was everywhere, tree branches and other flotsam left high and dry, a wire fence pushed over by the force of the current, and fresh silt patterning the ground. February had been a tumultuous month of rain, storm after storm, and weather system after weather system, unleashing its torrential fury and raising the river to unusually high levels, and spilling across the haugh on several occasions, creating a vast inland sea.

Now, at long last, the waters were receding, and I had clad myself in chest waders to enable me to explore the resultant mosaic of pools. The nearby river was still high, but nonetheless slowly losing its air of menace. I stopped by a large pool, water seeping out of it and creating its own stream that wound its way back to the main river.

The footprints of a heron pockmarked silt deposited by the flood, and by a rush-filled margin, a snipe flushed out by my feet, taking to the air in a flurry of wings and whirring away high into the distance. But frogs were my focus, and truth be told, I've been a frog addict from my earliest days – a boy thing, I suppose, and when growing up in Edinburgh I would eagerly seek out the frog-spawning ditches

Snipe

by disused railway embankments. The best time to witness these mating congregations was in mid-morning when the sun was shining – the frogs loved the warmth it brought which spurred them into activity.

That is only half the story, however, because frog-spawning seems to be split into two distinct strategies. In the former, the frogs can be around for a week or so in the breeding ponds and ditches, often croaking when it's warm, whereas at other times they spawn at the dead of night and are gone the next day. This latter approach to breeding is the most frequent, so when it comes to watching frogs, it is all a bit of a lottery.

As I splashed across the Devon flood plain, there was no sign of any frogs in the temporary pools, but I had not expected to find any in such places, for frogs typically have their favoured sites for spawning. I knew of one such shallow pond nearby where frogs spawn year after year, so I headed cautiously towards it. The sun had emerged, too, which was a good sign and would hopefully provide the frogs with some encouragement to stir.

Once I was reasonably close to the water's edge, I stooped down, and then ever so slowly, and an inch at a time, crawled across the damp, muddy ground towards the pool. When a few feet away, I could see a small shiny protrusion above the limey-green surface film of floating duckweed – a frog's head.

Frogs are typically wary creatures, especially during the day, and I knew that one sudden movement on my part would send this little amphibian swirling down to the bottom of the pool. With a bit more pulling of the arms and pushing of the legs, I slithered my way to the fringe of this wild pond. The beady eyes of the frog stared at me and I stared back. It soon decided I represented no danger and it began to gently croak.

In frog-speak, these croaks were probably a thing of great musical beauty, and even to the human ear they have an appealing resonance. The frog's pale throat-patch quivered in croaking contentment, and then another head popped up. The croaking turned into an amphibian harmony, a real-life version of the frog chorus.

Frogs

It is hard to describe the wave of happiness that swept upon me as I lay spread-eagled on the ground right next to this duo of croaking frogs. At that moment, there was nowhere else in the world I would have rather been. The power of nature to relax and destress the mind is inestimable. Nature is a great soother, a miraculous medicine that lies in boundless quantities on our very doorsteps.

Knowing that frogs were now returning to their breeding ponds, the following evening I walked a couple of lanes near the river with torch in hand in the hope of spotting some on the move. They are much easier to spot when caught in the openness of crossing small roads than when in among grass and other vegetation. The conditions were perfect: light drizzle and the air relatively mild, despite it being early March. I soon spotted the familiar form of a frog crossing the lane, and then several more, their pale throats catching the torchlight.

For some reason, frogs always seem bolder and less shy at night-time, and these ones were totally unconcerned by my near approach. This could be because they are so consumed by the urge to reach their breeding ponds, or perhaps the torchlight confused and dazzled them. Whatever the case, when I crouched down to examine one, it made no attempt to retreat and happily stood its ground. Several others resting in a nearby lane-side puddle were equally bold. I also spotted several toads, which was most unusual, as they usually do not emerge from hibernation until two or three weeks after frogs.

The timing of frog-spawning has always been of great interest to me, because it is an indicator of our changing climate and global warming. As a boy, back in the 1970s, it was in the middle of March when I would find the first spawn in central Scotland. Now it can be as early as the end of February. That shift in days might not seem much, but in nature terms it is gargantuan.

Of course, geography is a major dictator of spawning times. The pioneering nature diarist, Gilbert White, in *The Natural History and Antiquities of Selborne*, published in 1789, recorded 28 February as the earliest date he found frog spawn. I would imagine the first spawning date in Hampshire is now very much earlier than that.

This contrasts sharply with the high ground in the Cairngorms where I have come across fresh frog spawn in mountain pools in late April, as well as palmate newt tadpoles that had not even metamorphosed into adults in their first year and were taking a second summer to finish the job off.

Palmate Newt

Later in the week, I returned to the frog pool. There were no signs of any frogs, but they had left behind several clumps of jellified spawn, which despite the dull light, glinted like shining beacons. The frogs had completed their task and now they will disperse to spend the summer fattening up on the abundance of slugs and other small invertebrates found on the haugh.

I felt a tinge of nostalgia too, for I adore finding frog spawn and it is one of the keystones of my childhood. As such, these incredible black-specked spheres of jelly are a wonderful reminder of my first early connections with nature.

Chapter 19
SPRING BY THE RIVER
(March)

A spring dawn-frosted morning on the Devon just a couple of days before the March 2020 'Coronavirus Lockdown'; still air, azure sky and sunbeams brimming over the rolling horizon, spilling forth a myriad of sparkling rays.

Nature is so inspiring, life-giving and powerful in every way, and here by the river it was unfurling its beauty in such a spellbinding manner that tears welled up in my eyes. Naturally, my emotions were partly stirred by the immense challenge humankind was facing, but in a strange way that was a positive, focusing the mind on what a wonderful world we live in.

It also brought contemplation on my perception of the natural world and how it has changed over time. When I was younger my brain was more scientific in manner; nature being something to research and study. Why does a fox do this, or a lizard that? Such an approach is, without doubt, important, because the more we know about nature, then the better we can protect it. But as the years have progressed, my mind has also become more reflective; rather than knowing why, for me, it is much better to enjoy.

I wandered down to my favourite part of the middle river. There were signs of spring everywhere: singing birds, frog spawn in a nearby frozen-mirrored pool, and silver-furred catkins adorning the riverside willows. On the top of a high alder, a male song thrush, with his pale-speckled breast catching the soft sunlight, sang his little heart out, a sweet melody of ringing notes, so true and sweet. Not to be outdone, down in among the tangled roots of a riverside alder, a diminutive wren shivered in the sheer passion of delivering magical music.

In the distance by the flood meadow, the wonderful liquid trilling of a curlew drifted across the breeze – such a

Curlew

beautiful and haunting sound. Nature was busy at work, and life felt good.

Then, something remarkable happened. It was just a glimmer, a chance discovery and no more than that: a smooth mossy dome in the fork of an elder. I could have walked past it a thousand times and not seen it, such was the way it seamlessly blended with the branches. This domed marvel was the nest of a long-tailed tit – an intricate engineering masterpiece woven from moss, lichen and cobwebs, and lined with hundreds of feathers to keep it snug.

Inside, a female long-tailed tit, with her tail kinked over her back, was incubating her clutch of eggs, safely cocooned in her near-invisible nest.

Nearby, and out of sight in bramble thickets and hedgerow tangles, blackbirds, song thrushes and other birds would also be sitting on their own nests, nurturing and providing warmth to their fragile eggs.

Such imagery was wonderfully heart-lifting; a whole new generation was on the cusp of hatching, bringing new vibrancy and wonder to our everyday lives.

Blackbird

I meandered further on, immersed in the soft sunlight filtering its way down through towering alders and which glinted upon a drift of yellow and white polka dot flowers that peppered the river bankside. The yellow flowers were lesser celandines and the white, wood anemones; two of our earliest flowering wild plants. They were such beauties: gold and snowy-white spangled colour, reflecting hope and promise in this season of renewal.

Both plants are very much woodland specialists, and their presence is often an indication that the ground on which they grow is long undisturbed ancient forest, which unfolds a special facet about rivers. Wood anemones are

Wood Anemone

especially good ancient woodland indicators because they spread so very slowly by means of swollen roots called rhizomes, which creep through the soil.

In many instances, river banksides have been left untamed and wild for generations, and consequently, they are creeping tendrils of ancient wildwood, forming a vital wildlife habitat.

In spring, it is the wildflowers that make these places so very special. The early flowering of lesser celandines and wood anemones benefits nectar-seeking insects at a time when there are precious few other wildflowers about. For early emerging bumblebees, a sun-glowed clump of wildflowers on the riverbank is an oasis, a provider of life.

It makes sense to bloom in early spring, because the surrounding trees will soon envelop into full leaf, while later, taller-flowering plants such as cow parsley and meadowsweet will also predominate, resulting in the ground becoming a darker and more shaded place as spring turns into summer.

Bumblebee

I knelt upon the ground to examine one of the flowering patches of wood anemones on the river bankside. A gentle breeze momentarily took hold and the little white petals of the flowers quivered and shook in quite delightful fashion. How appropriate, I thought, for the wood anemone is often known as the windflower and, indeed, is believed to be so named after the Greek word for wind, *anemos*.

In Greek mythology, anemone flowers sprang up where Aphrodite's tears fell as she wept over the death of her lover, Adonis. Other folklore connects the anemone to magical fairies, who were believed to sleep within their petals after they closed at sunset.

Wood anemones and lesser celandines are flowering jewels that have inspired humankind from the earliest of times – and they continue to do so to this day. The nineteenth-century poet John Clare wrote of wood anemones as being "weeping flowers in thousands pearled in dew" ('Wood Anemone', *n.d.*). A perfect description for such perfect flowers.

However, there needs to be sunshine aplenty for one to enjoy both these spring flowers in their full glory, for when it is rainy or cold their outlook is one of contrasting dreariness, with the petals of the multitude of brilliant flowers having closed-up like a thousand clenched fists. Or as William Wordsworth observed:

> *There is a flower, the lesser celandine*
> *That shrinks, like many more, from cold and rain*
> *And, the first moment the sun may shine*
> *Bright as the sun himself, 'tis out again!*

William Wordsworth, 'The Small Celandine' (1804)

The celandine closes for a reason: dry pollen is much easier for insects to transfer and shutting the petals helps prevent dampness within the flower.

Other flowers were emerging too, and as I slowly walked along the riverbank the ground beneath my feet was undergoing a remarkable transformation. The verdant green, elongated leaves of wild garlic, or ramsons as they are known, were unfurling and soon their blousy dome-shaped white blooms would burst into flower.

An old country name for ramsons is 'stinking nanny', given its garlicky aroma – but I'm dubious about the aptness of this term, for I find the garlic smell as being subtle and having an aura that is most compelling. Ramsons can form very dense colonies, but they are fickle in their distribution, being abundant in some areas and totally absent in others.

I headed away from the main river and followed a tributary burn up into nearby woodland. Here, by the steep banks of the burn, splashes of lemon and limey-green hues caught my eye, a citrus carpet of flowing colour. It was a creeping mat of opposite-leaved golden saxifrage, a real tongue-twister of a name, but a marvellous and rather understated spring flower for all that. Their golden-green flowers are truly miniscule but so exquisite and intricate in their form.

Another bright glow of colour shone out ahead of me on a small island in the burn. It was a clump of primroses just emerging into flower. Yellow

Wild Garlic, 'Ramsons'

Wild Primrose

comes in all shades and tints, but there is surely no shade of yellow that compares in brightness and vibrancy as that found on our wild primroses.

Primroses deliver such dazzling freshness, little orbs of joyous sunshine. Primrose – *prima rosa* – the first rose or flower of the year, and one which for generations have been picked as Easter decorations for churches across the land.

I ventured back down to the riverside and on the inside cut of a meander, silvery pussy willow catkins adorned swaying branches like Christmas baubles. As is the case with the alder, willows are well adapted for living by riversides and in semi-submerged environments. Identifying different types of willow can be challenging, as they often hybridise with each other, but on the Devon the most prevalent species are grey willow, goat willow and crack willow. Willows are ecologically crucial, pioneering colonisers, riverbank binders, and a key food plant for larval and adult insects.

The next day, I spent the whole afternoon down on the floodplain, walking from Dollar to Tillicoultry and then back again. It was a glorious day, soft spring sunshine and barely a whisper of a wind. However, a storm was gathering for humanity in the form of coronavirus and I sensed things were about to change in the available time I could spend down by the river. The popular walkway on the far side of the Devon was also much busier than normal, indicating the looming turmoil. With non-essential shops and other public places beginning to close their doors and mass gatherings advised against, people were returning to nature in their droves, and on that wonderful spring afternoon, there was no better place to be.

This period of spring, starting from mid-March and stretching into early April is one of overlap and transformation, and I felt privileged to be in among both winter and summer wildlife at the same time. On a flooded margin, winter-visiting greylag geese slumbered and gently honked to each other, while a restless flock of fieldfares cackled as they bounded through the bare branches of a stand of oaks. Such lingering signs of winter were still strong, but it was yielding and soon the greylags

Fieldfares

and fieldfares would depart to their northern breeding grounds. Spring now had the upper hand and earlier that day I had already heard a singing chiffchaff. They are the first of our warblers to arrive, and some people find their two-tone 'chiff-chaff' call to be rather mono-tonous. It is not an opinion I share, for the simple song has a reassuring quality, and when other warblers have fallen silent in August, the little chiffchaff will continue to sing, albeit not as enthusiastically as before.

By each turn of the river, I was making new discoveries to enthral and inspire me, including a pair of sand martins flitting over the water, my first ones of the year and newly arrived from their African wintering grounds. They brought a real sense of joy and anticipation, with memories drifting back into my mind of the two

Chiffchaff

young bedraggled and frightened sand martins I had found the previous summer that had so hastily fled their flood-struck nesting burrow.

There was still a prolific abundance of pools dotted across the riverside meadows, a legacy of the wet winter just past. A high-pitched 'pee-wee' pierced the air; a wonderful, evocative call, and one that stirred memories of wild camping in highland glens in spring. It was a lapwing, a stunning iridescent green plover with a distinctive crest. This one was a male and he suddenly took to the air from the edge of a large pool and engaged in a series of stunning aerial manoeuvres as he embarked upon his courtship display.

Calling excitedly, the lapwing rose steeply on slow and measured wing beats, swept around in a wide circle, gained height again and then, zoom!

Down he plunged and rolled, twisting this way and that, his broad wings all over the place and somewhat akin to a crazy kite spiralling out of control. But this lapwing knew exactly what he was doing and at the last moment, when a crash landing seemed inevitable, he stalled and alighted on the ground. A female must have been somewhere near, hiding in the rushes.

Lapwing

On previous occasions when watching such aerial displays, I've been so close to courting males that the air hummed and throbbed from the noise of their energetic wing beats.

There were other signs of spring by the riverside. Queen buff-tailed bumblebees buzzed and hovered over the ground, looking for holes or crevices to settle down and form new colonies. The previous autumn all the workers, drones and the old queens died off as the cold weather took grip, leaving behind just the young fertilised queens, who hibernate, before emerging in spring to start the cycle of life once more.

By the edge of a nearby field, black-thorns were frosted with white-dazzled flowers. Blackthorns flower first, with the leaves following later. Even when the weather turns freezing the flowers will still blossom, with a cold spring traditionally being known as a 'black-thorn winter'. The blackthorn is best known for its bitter tasting fruits we all know most familiarly as sloes, which are used for making wine, jam, and for flavouring gin.

Blackthorn

These blackthorn flowers were so exquisite, and I could not help but linger for a while to appreciate their natural elegance. Everything seemed so perfect that spring afternoon; nature was showing itself at its best, and as I looked up once more across the river to the throngs of people on the distant walkway, it was apparent that many others were also enjoying its wondrous beauty.

Chapter 20
A DAWN SURPRISE
(March)

On the dawn of the day when the 'Coronavirus Lockdown' was announced, I ventured down to the lower Devon to retrieve my trail camera, realising that in the weeks to come my movements were likely to be restricted.

Earlier that month, I had noticed plenty of beaver signs along this part of the river, most notably stripped bark from fallen tree branches lying in the river. These were not beaver-felled branches; they had tumbled into the river of their own accord. It seemed that the beavers here were patrolling the river edge and feeding on the soft inner bark and twigs of these branches without having to leave the water. Bark is a key foodstuff of beavers, especially in winter. In spring and summer, aquatic plants, roots, leaves and other vegetation become important elements of their diet.

One freshly stripped branch looked ideal for video monitoring, so I secured my trail camera to a nearby tree stump, which I planned to leave for several days. I had been keen to get my camera down to this part of the river for weeks, but the perpetual rain had made that a risky proposition with the water being very high and likely to rise even more so at any time and quickly submerge the device.

The inevitable happened, and the day after placing my camera, it rained heavily once more and when I returned to try and rescue it, I was too late, for it was already a couple of feet under the surging water. Reluctant to lose the camera, I stripped off and waded into the icy water and with numb fingers, gingerly felt my way down the stump and managed to unstrap it after some fumbling. The water was perishing because it was still partially being fuelled by snow meltwater draining down from the Ochils, so it was a relief to finally flop back onto the bankside, cold and shivering, and get my clothes back on. At least I had retrieved the camera and hopefully the memory card would still be salvageable.

It is a perpetual hazard placing trail cameras by the river's edge, and over the course of my river year, I had already lost two cameras. Fortunately, they do sometimes survive after being submerged, and despite having been under the water for several hours, this one was still functioning. On examining the video contents, there was no beaver footage, but there were several clips of moorhens balancing deftly on the fallen branch.

I had never previously noticed moorhens being abundant on the river, but this year they had been especially so. Here on the Devon, they are incredibly shy birds, scuttling for bankside cover at the first hint of an approaching person. This is in stark contrast to moorhens that live in park ponds and lakes, which are tame and trusting.

Once the water had fallen back again by mid-March, I sited the camera back at the exact same location by the bark-stripped beaver branch. A few days later, the coronavirus pandemic was gathering pace, and with the country about to enter lockdown, I had no option other than to retrieve the camera whilst I still could.

Dawn had just broken, and it was one of those marvellous mornings where nature brimmed over in its spring abundance. Skeins of greylag geese honked their way across the sky and a kingfisher flashed up the river as soon as I had reached the bankside. Then, a movement in the water – a brown furry head, moving upstream about 50 yards away from me. An otter, I thought – but no, the head was broader, and the fur coloration was not quite right, as it had a warmer tone to it. Excitement coursed through my veins – it was a beaver, and the first wild one I had ever seen.

Beaver

Keeping low, I carefully followed the beaver as it swam upstream, its broad paddle-tail sometimes breaking the surface as it rippled forth into a new burst of propulsion. I was pretty sure it had seen me, but for the moment it was just keeping a wary eye on me. I retreated from the bank, quickly trotted upstream, and then slowly trod back to the riverside once more in the hope of heading it off and getting a closer view. As my head popped back over the bankside, I managed to take a quick couple of photographs, but the beaver immediately saw me and dived under with an impressive tail-thumping splash. I scolded myself. It was not my intention to alarm the beaver, but enthusiasm had got the better of me.

Such spectacular splashes from their paddle-tail when diving under is typical beaver behaviour when frightened, and acts as an alarm mechanism to alert other beavers. I waited by the bankside for several more minutes, but there was no more sign of the animal. It could have swum underwater for several hundred yards before surfacing, or perhaps it had fled to the underwater entrance of its secret bankside lodge. Certainly, beavers can stay under water for several minutes, and once submerged, have a whole range of distancing options available, both for travelling upstream and downstream, and for seeking shelter in the thick tangle of overhanging bankside vegetation.

Whatever escape route this one took, I was thrilled to the core at the sighting, although an air of disappointment hung over me at my lack of stealth when stalking the beaver. Of all people, I should have known better.

It also highlighted that when it comes to wildlife watching, chance is more times than not the underpinning factor. With my intention for a planned programme of spring dawn and dusk watching for otters and beavers at various vantage points along the river now in disarray due to looming coronavirus restrictions, Lady Luck had come to the rescue.

I wandered several hundred yards further upstream, my mind still buzzing with excitement from the beaver encounter, and then retrieved the trail camera.

On returning home, the camera contents were wonderful: a plethora of night-time videos of two different feeding beavers, one slightly larger than the other. They never appeared on film together, but the size difference was discernible. Beavers live in small family groups and youngsters can stay with their parents for up to two years before leaving to create their own territories. The first video clip showed the larger of the two beavers swimming ever so slowly past the branch, almost as if in quiet contemplation. It was difficult to be sure, but it was possible that it had seen the camera and was giving it a casual once-over.

Other film clips revealed the beavers feeding on the inner layers of bark or on twigs, either when in the water, or at other times after they had climbed onto the bough, looking like giant hunched hamsters as they perched and gnawed away with their powerful teeth.

For me, it was a ground-breaking moment and I was at long last beginning to get to know my local beavers. With their penchant for feeding upon bark from autumn through to early spring, beavers prefer to eat softer-wood trees such as willow, hazel and aspen, and avoid harder species like alder.

Having the ability to eat twigs and the soft inner bark of trees is a quite remarkable evolutionary dietary adaptation. A long appendix aids a beaver in digesting this high cellulose bark diet, with a cocktail of micro-organisms in the gut further helping in the breakdown of woody material to aid absorption of nutrients by the gut.

Fungi are recognised as one of the few groups of organisms able to break down wood, thus returning valuable nutrients back to the soil. As I pondered the videos, it struck me that beavers essentially do the same, and along with the many other ecological benefits they provide, they are key recyclers of trees, many of which have fallen and perished after gales or from the pressure of flood torrents. By unlocking and returning nutrients back to the environment, beavers are ensuring that there is drive and impetus behind the Devon's cycle of life.

Chapter 21
GETTING TO GRIPS WITH LAMPREYS
(April)

A sweet note, and then one more: a musical couplet, followed immediately by another, hesitant at first but gaining in confidence as the music from a song thrush floated across the crisp dawn air.

Song Thrush

Each April morning as I walked the same circuit by the middle river, this song thrush had become a familiar companion, always perched on its same 'song post' atop a riverbank alder. The range and variety of the song was immense, with each phrase repeated before moving onto the next. It was a song thrush in full flow at dusk that inspired Thomas Hardy to write of a bird that had "chosen to fling his soul upon the growing gloom" with an "ecstatic sound" of "joy unlimited" ('The Darkling Thrush', (1900)).

Due to the coronavirus restrictions, my time down by the Devon was limited to one short exercise visit per day on a small stretch of the middle river close to my home, with other parts being out of bounds for being too far away. No matter, for it brought an opportunity to focus in detail on a single part of the Devon, and this short daily circuit along one bank, across a bridge, and then back along the other side revealed a bounty of wonderful natural discoveries as the month progressed.

Much of April brought marvellous spells of spring weather with many of the dawns cold and frosty, followed by the sun brimming over the horizon for the rest of the day, bringing real warmth to the air. Great-spotted woodpeckers drummed on trees, chiffchaffs and willow warblers sang their little hearts out and rooks busily picked over the recently sown fields by the riverside – nature thrived while humanity suffered.

My walking route took me past my favourite backwash pool, where one morning I watched a young heron stalk the shallows and pounce upon a small sliver of a fish, which I presumed was a diminutive eel. I posted a photo of the heron with its wriggling catch on social media, and one observer soon messaged back saying he was certain the fish was in fact a brook lamprey. I examined the photo more closely, and yes, he was right, for this thread-like fish had different bodily proportions from an eel, being thinner and slightly shorter.

Brook lampreys spawn in the spring, and now alerted to the fact that they were out and about on the Devon, I ventured down to the backwash pool the following dawn, carefully searching the shallow water margins in the hope of spotting one. Astonishingly, when I peered over the bank, I found numerous spawning toads instead.

Common Toad

Toads undertake their breeding with zeal and enthusiasm that borders on the unhealthy, and here there were writhing masses of the warty amphibians in the languid shallow water by the bankside. Many of the toads were in the mating position known as 'amplexus' (the Latin for 'embrace') where the males grip tenaciously to backs of females in readiness to fertilise the string of eggs when released into the water.

Toad mating is enough to put even the most depraved Roman orgy to shame and it is a real wonder where these toads managed to pull their energy from, especially since they were thin and emaciated after having just emerged from hibernation. Such is the incredible intensity of the occasion that for many toads this will be their final act and they will perish from exhaustion. In one shallow pool by the edge of willow carr, I came across such a mortal result where there was a threesome of toads in a ball, with the animal at the bottom having been killed by the vice-like grip delivered by one of the others. In a nearby muddy inlet, the bodies of two other toads lay on the bottom.

The most remarkable aspect to this toad behaviour is that they were spawning in the river, which is something I have never seen before, as typically they breed in large ponds and lochans. Although this backwash pool is calm and placid like a pond, it is connected directly to the open river by a channel several feet wide and is exposed, at times, to the vagaries of the Devon's powerful flow. The toads had probably successfully spawned here for many years, as they tend to be traditionalists in their breeding sites, but I did wonder how the tadpoles would cope if the river turned to spate as it so frequently did, even in spring and summer. The only conclusion I could reach was that they do survive such events, perhaps by burrowing into the sediment for protection from the surging current. Compared to the relatively benign environment of an enclosed pond, the tadpoles would also have to contend with predation by trout, kingfishers, and dippers.

Thrilled with the toad discovery, I wandered further along the bankside scanning the water. Then, I glimpsed an undulatory movement in the mud where the water was no more than an inch deep. A brook lamprey! Letting out a yelp of excitement, I clambered down the bank and into the water, my wellies sinking so deep into the mud that they began to fill with a cold trickle.

The lamprey, which was no more than five inches long and as thin as a piece of cord, was cornered in the muddy shallows and I gently scooped it out the water with cupped hands. I opened my hands to examine it, but as quick as a flash it slipped through my fingers and back into the water. I managed to catch it once more, but again it wriggled free, such was its slipperiness. It reminded me of trying to catch butterfish in rockpools down by the coast, which are also the masters of escapology by squeezing through the smallest of gaps between fingers.

The lamprey disappeared into one of my muddy footprints in the riverbed where there was no chance of ever finding and recapturing it again. I climbed back up the bank and spotted another lamprey wriggling gently in the shallows, but it was too far away, and the mud too soft for me to wade in for a closer look.

Brook lampreys are most intriguing eel-like fish and the only time I had ever come across them before was as a teenager when searching a shallow gravelly margin of the River North Esk near Penicuik in Midlothian. Several of them had gathered together in a spawning frenzy, and I remember watching spellbound as they writhed and wriggled, so consumed by the act of mating that they were totally oblivious to my presence. Interestingly, I also recall that toads were spawning in a nearby pond, underlining that lamprey emergence probably reaches its peak in early April.

Most species of lamprey are migratory and spend part of their life cycle in the sea. They also represent some of the most primitive vertebrates alive today, being over 360 million years old. Technically, they are not true fish at all since they possess no jaws and have only a cartilaginous skeleton. They live for several years as larvae in streams and lakes, burrowed in silt beds, where they filter feed on micro-organisms. In this juvenile stage they are known as ammocoetes.

River Lamprey

Adult lamprey species have a toothed sucker mouth used to latch onto other fish to feed upon their bodily fluids. The diminutive brook lamprey, however, is different as it spends its whole life in a river, and after several years in the mud as an ammocoete, it metamorphoses into a non-feeding adult, which then spawns in groups on gravel or sand beds before dying. I imagine that the adult brook lampreys I had discovered had just emerged from their ammocoete larval stage in the mud in the backwash pool and would soon migrate, or drift downstream, to a nearby part of the river to spawn where the bottom substrate was more suitable.

The Devon also holds river lampreys and several years ago I found a dead one on the middle stretch of the river, which had succumbed after spawning. As adults, rivers lampreys are much larger (typically 12 inches in length) than brook lampreys, and after spending four or five years as larvae in the river, they migrate downstream to estuarine and coastal waters to parasitise on other fish such as flounders, before returning to the river in spring to spawn. During these river migrations, adult river lampreys move upstream at night and rest under cover during the day.

One bizarre aspect of river and brook lampreys is that they may be the same species, and just as how sea trout and brown trout differ in behaviour yet are really one and the same, these two types of lamprey have evolved their own different ecological lifestyles. The brook lampreys on the Devon were a fascinating discovery and as I headed for home, I was grateful at having had the opportunity to observe at such close quarters one on the river's most mysterious creatures.

It also confirmed in my mind how important these placid side pools are to the ecology of our rivers. This pool that I had come to know so well over the months just kept on revealing surprises, and the miracle of the larval ammocoete lampreys living deep within its silty bounds was one of the most remarkable yet.

By mid-April, the first of the sandpipers had arrived on the river from their wintering grounds in west Africa, making their presence felt with their distinctive trilling calls. They are delightful little greyish waders that especially favour shingle banks, frequently bobbing the head and tail, and when one takes flight, it will skim low over the water, wings drooping whenever gliding. These first arrivals were on migration and just passing through, with the Devon's main influx of breeding sandpipers arriving a week or two later.

The elaborate courtship ritual of sandpipers is always a joy to watch. The male often runs behind the female with fluttering wings, and he will also pursue her in the air, following an erratic course with deliberate and

Common Sandpiper

rather stiff wing beats. Sandpipers are fleeting visitors, only on the Devon for a short period of a few months with the majority having left again by early August as soon as they have raised their chicks.

Another April dawn on the Devon, and just before I had reached the bridge spanning the river on my last leg home, there was a tremendous splash in the water on the near bankside only a few feet away from me. I rushed over to the bank edge, and through a gap between two alder trunks, bubbles rippled across the water's surface. Only two animals could have caused such a splash – an otter or a beaver.

I tried to outguess the animal. If it were me making the escape, I reasoned, I would swim underwater downstream because not only would that be easier through being aided by the water's flow, but also the river was deeper there. Besides, further upstream lay a shallow riffling rapid, which would force any creature to emerge from the water and reveal itself.

I trotted downstream as gently as I dared, and then waited, hoping for a furry head to emerge on the surface beside me. Nothing. I looked back upstream and by the riffle I could see a pair of goosanders. Could it have been the goosanders that had caused the splash and then moved quickly upstream? Unlikely, but possible, I supposed. Then, close to the goosanders, a round head bobbed up in the pool above the riffle and moved slowly across to the opposite bank. The buoyant nature in which the head and body were carried in the water told me immediately it was a beaver, so I fast-walked back upstream, mindful of how indiscretion on my first beaver encounter had panicked the animal.

Soon, I was directly opposite the beaver as it slowly headed upstream by the far bank, the animal keeping a close eye on me all the while as I took some photographs. When up close against the bank, this beaver was remarkably hard to discern, blending in harmony with the tangle of alder roots, and I suspect many people on riverside walks must walk past beavers without realising they are even there. This beaver swam under the bridge and then gently submerged, leaving behind a shimmer of bubbles, after which I lost it for good.

This was the first indication I had ever had of beavers occurring so far upstream on the Devon, and I pondered whether this might have been a two-year-old animal that had left its family natal territory further downstream and was now investigating pastures new, possibly in an attempt to establish its own territory. Certainly, it hung around the area for a while, as later that week I found two felled willow saplings by the backwash pool, the bark neatly stripped.

After spending so much time over the previous months trying to spot beavers, I had now seen two in the space of only a few weeks. Along with the lampreys, toads, and sandpipers, April was turning out to be a magical month on the Devon, with many more of its secrets unfolding before my eyes, despite my attention being drawn to only one small section of the river.

Chapter 22
SPIRITS OF THE NIGHT
(April)

With my movements restricted due to the 'Coronavirus Lockdown' measures, I had found another option to explore the Devon catchment that lay on my very doorstep, for I was fortunate that one of its tributary burns runs its course by the bottom of my garden.

On many evenings during April, I would slip through a gate in the garden fence and take a handful of steps down a wooded slope to sit by the burn's bubbling course, a serene place where just the peace of nature took hold, enabling my mind to drift in quiet contemplation. This burn was just as much a part of the Devon as its main winding course, and it too, had much wildlife to reveal.

During one of these reflective evenings, I wondered how the burn's invertebrate life compared to the main river, so the following day I returned with a small hand-held sweep net and a plastic tray. Dippers haunt the burn in autumn and winter, but I had never known them to breed here, presumably due to a scarcity of food.

On sampling the burn, I could not have been more wrong, because there was in fact a large variety of larval invertebrate life, including caddis larvae in their gravelly pupation shelters and numerous mayfly and stonefly nymphs crawling on the under-surfaces of stones. Also captured were freshwater shrimps (amphipods) and the tiny thread-like wriggling forms of midge larvae. The burn was too shallow to 'kick-sample', so instead I scrabbled my fingers over the gravel and pebbles, enabling this abundance of tiny life to float into my net. As I was doing so, a pair of grey wagtails swept up the burn in an undulating flight and over my head.

If there was ever a misnomer, then surely it must apply to the grey wagtail, for rather than being drab as the name suggests, it is in fact one of our most colourful birds, especially the male with his most striking yellow underparts. The grey wagtail is a specialist of fast-flowing rivers and streams where it snaps up insects and other invertebrates along the water's edge, rather than being a specialist underwater forager like the dipper.

One feeding method wagtails often use is to perch on a rock in the middle of the burn, and then with a short flutter, catch a fly in the air before settling back onto the lookout post.

Grey wagtails frequently hang around in pairs and have a distinctive 'chirrup' call. Although they are quite flighty birds when disturbed, they will soon alight again, often on a boulder in the middle of the stream. I suspected this pair had a nest secreted in a nearby bankside recess, as I had seen them regularly on the burn over previous weeks.

As I examined my collection tray for invertebrates, I noticed out of the corner of my eye several adult mayflies in flight nearby, dancing up and down just above the surface of the burn. They were a type of mayfly known by anglers as the 'large dark olive' and are typically on the wing in March and April, although most usually they hatch during the middle of the day, rather than the evening as was happening here. That day, the air was mild, and perhaps that had encouraged them to emerge.

They were wonderful to watch, an aerial jig above the tumble of the burn. Where there are flying insects at dusk, then there are also likely to be bats. I had already planned to watch bats that evening and had brought with me an electronic bat detector, and there was every possibility that these hatching mayflies would prove attractive to bats. The bat detector is small, not much bigger than a mobile phone; it converts the high frequency ultrasonic calls of bats into sounds that can be discerned by the human ear. Bats locate their insect prey and objects in the environment by the pattern of returning echoes from their calls – a system called echolocation.

After having finished my invertebrate investigation, and with dusk now taking hold, I sat by the edge of the burn, slipped my bat detector earphones on, and waited. At first, only the soft warble of a nearby robin settling down for the night breezed through the air. Then, high in the tree canopy, a bat on flashing wings swooped and swerved, and in the fade of the gloaming my headphones suddenly crackled into life, filling my ears with a cacophony of weird staccato clicks that gained rapidly in intensity before drifting away again as the bat spirited away.

The bat soon returned, filling more rapid-fire chatters into my earphones. These echolocation calls are a most amazing adaptation, and I was consumed with awe at the forces of evolution which developed such a remarkable hunting system. For bats, it is their eyes in the dark, and the echolocation is so effective in painting a detailed picture of the environment, that the bat I was watching weaved and twisted its way through the tangle of tree branches without the slightest hesitation.

Bat detectors are useful for naturalists because as well as giving an insight into bats' secret ultrasonic world, they also aid in species identification. This is because each type of bat generally has a distinctive frequency call range. The peak frequency of the bat flying above the burn was 55 kHz, which told me it was most probably a soprano pipistrelle, a bat that has an affinity with aquatic habitats. In my garden, just a stone's throw away, I often hear the closely related common pipistrelle, which has a peak call frequency of around 45 kHz.

Daubenton's Bat

I knew from bat detecting visits on previous occasions that along the main river section the Daubenton's bat – sometimes known as the water bat – is the most prevalent species. Its flight pattern is so distinctive compared to other bats that I did not even require an echolocation detector to make a positive identification as they skim low over the river's surface on fast-beating wings, which are often held high, presumably to avoid contact with the water.

This specialist hunting technique enables them to target a range of small insects, which are often scooped from the water surface by using their tail membrane and feet, with the bat then rolling forward to gobble its prey while still in flight. When fly-fishing on the Devon at dusk, it is not unusual to see a Daubenton's bat swoop towards the fly cast as it unfurls through the air, momentarily attracted by it. On several occasions, I have even felt the distinctive vibration of a bat's wings brushing against the fly-line, which I suspect they did deliberately to try and get a sense of feel and perspective as to what was the new strange object in their environment.

Sitting by this small burn by my house became something of an evening ritual and I enjoyed the quiet solitude of its calm enfoldment. Over the millennia, this burn had incised a miniature valley – in some parts of Scotland often known as a dean or den – where a host of trees and wildflowers flourished, protected from urban development or agriculture. It is a magnificent narrow ribbon of wildness stretching from the Devon up into the Ochils: a place where bats and a multitude of other creatures can thrive and prosper.

Chapter 23
THE HIDDEN RICHES OF THE ESTUARY
(May–June)

Towards the end of May and into early June, I concentrated the final spell of my year-long river journey on the Devon estuary by the village of Cambus, mindful that this part of the river still had many more secrets to reveal.

After several weeks focused on the middle river because of the 'Coronavirus Lockdown' restrictions, it was like a breath of fresh air to be back on the estuary, with its open horizons and different selection of wildlife. Within minutes of my first arrival, I had stumbled upon a young cormorant resting on a rock in the middle of the estuary and a pair of shelducks dabbling about in the shallows. So very different from the middle river and wonderful to see.

Both were interesting finds because they provided a good indication of the abundance of other life in the estuary, with cormorants being fish feeders, and shelducks avid devourers of tiny invertebrates. The feeding behaviour of these shelducks reminded me

Shelducks

of flamingos, sweeping their bowed heads from side-to-side in the shallows and using their especially designed bills to filter diminutive creatures from the glutinous ooze.

In winter, I often see shelduck in the ephemeral pools on the far side of the estuary, and I wondered whether these ducks I had just discovered were breeding birds. Shelduck nest in holes and hollows, but are not much good at digging, so will often take up residence in rabbit burrows. It makes real sense – after all, why go to the effort of digging when some other creature can do the work for you? Shelducks are chunky birds and it must be a tight squeeze for a mother duck inside a rabbit burrow, although I imagine the female will do some remedial work to ensure the tunnel is fit for purpose.

I was keen to explore the underwater life of the estuary, despite the thick, soft mud and algae-covered rocks looking as uninviting as they had done back in the winter. My biggest problem was to find a route into the water where I would not become stuck in the deep sediment, and to find a stretch where the water visibility was clear enough to permit even some limited snorkelling.

I walked to the mouth of the estuary and noticed on the far bank a fan-shaped area of stones and small rocks leading down to the water's edge. I scrutinised it through my binoculars: this would provide a reasonable and stable platform for me to dig and turn over stones to seek invertebrates. However, the water here was as brown as a milky cup of coffee and not suitable for snorkelling. For that, I would need to try somewhere else.

I heard a moorhen calling from the adjacent reed-fringed Cambus Pools, and distracted by this new-found interest, I wandered over to one of these freshwater ponds and sat by the water's edge. It was a serenely peaceful place, with its open reflective water backdropped by the Ochils in the far distance. A drake mallard bobbed out on the water and sedge warblers engaged in their frenetic little song flights on high-held wings. Then, a movement in the reeds only a few feet away, and the most remarkable looking bird waded into view – a water rail!

I could probably count on the fingers of one hand the number of times I had seen a water rail before, for they are such elusive birds, living a secretive existence in reed beds and seldom exposing themselves to the outside world. About the size of a moorhen, with greyish-blue underparts and a curved red bill, water rails are probably commoner than one might

Water Rail

realise, and in previous years my trail cameras in riverside ditches had occasionally caught individuals on film.

This rail had seen me, but seemed totally unconcerned by my presence, possibly because it so rarely ever came into close contact with humankind and thus had no reason to fear me. Twice it buried its head under the water to probe for invertebrates, its eyes constantly scanning for the slightest movement of a creature for it to snap up. As quickly as it had emerged from the reeds, the mysterious rail paddled back into the thick cover and disappeared.

Buoyed by the sighting, I headed back to the estuary and then wandered upstream to reconnoitre for a place to snorkel. Soon, on the middle part of the estuary, I found a stretch that suited my needs. It was far from ideal, but with the tide low, a rock shelf had been revealed which I could crawl out onto, avoiding most of the mud. Below it was a short series of cascading pools, where the water visibility had reached the dizzy heights of a few inches – the best one could ever hope for in the estuary. At the bottom end of these pools was another rock lip, from where I could make my way back out of the water. I would return the following week and take the plunge.

A few days later, I revisited the estuary at low tide to investigate on foot the area of stone and rocks by the river mouth. Before venturing onto this rocky area, I sat close to the water's edge for a while, reflecting upon estuaries and their importance in the natural world. Estuaries are such unique places – an interface between freshwater and saltwater where rising and falling tides meet the river's flow. They are among our most dynamic and diverse habitats and their sheltered aspect and abundance of invertebrates, make them important nursery areas for fish and feeding grounds for waders and waterfowl.

A typical estuary is a two-way food conveyor belt with the river water flowing out to the sea and the infilling tide bringing marine richness back in. It is a natural smörgåsbord, a vast muddy acreage just oozing with life. Estuaries are places of wide empty skies that reinvigorate the mind and encapsulate the wonderful wildness of nature.

Only a few months previously, on a cold winter's day, I had visited the Eden Estuary near St Andrews in Fife and came away mesmerised by its abundance of wildlife. The mud that clawed at my boots was patterned with the empty half-shells of cockles and small clams called tellins, a sure sign of the bounteous life held within the glutinous gloop below. There were mud shrimps and worms in abundance in this inner section of the Eden estuary and so much other life too. The constant burbling of curlews carried far in the wind and on scanning the mudflats, I spotted oystercatchers, redshanks, dunlins, godwits, grey plovers and shelducks.

Oystercatchers

As I made my way along the edge of the Eden shoreline, I also glimpsed small groups of cormorants resting on mudbanks by the main channel of the river. As the tide flows in, flounders will nudge their way up the rapidly filling channels to feed upon the scores of molluscs, crustaceans, and worms. These flatfish will in turn be preyed upon by the cormorants and other fish-eating birds such as mergansers.

Sitting on a slimy, algae-covered rock on the Devon estuary, I instinctively knew there would be no such riches here. Much of this is down to the natural decline in biodiversity that occurs when salinity drops and the water becomes fresher, such as happens the further up the estuary of the Forth one goes. I also suspected there was a legacy of lingering pollution at work.

As was the case with the main

Cormorants

course of the Devon, pollution has had a major impact upon the invertebrate biodiversity of the Devon estuary at Cambus. Estuaries are especially susceptible to pollution because they are the endpoint from where contaminants from inland run down from the river to meet with seaborne pollution carried up by the tides. As river and tidal currents meet, they slow and buffer, resulting in contaminants being deposited in thick beds of mud.

Prior to coming down to the estuary, for research purposes, I had read a fascinating article by distinguished estuary biologist Donald S McLusky published in the annual periodical *The Forth Naturalist and Historian* (Volume 38, 2015).The article recounted how a scheme to drain the Carse of Stirling first embarked upon in the second half of the eighteenth century, which involved removing overlying peat bog, resulted in decades of massive sediment deposition in the inner Forth from Stirling to Bo'ness that destroyed the livelihoods of oyster gatherers and other fishermen in the area. This was further exacerbated over the next couple of centuries by pollution from the growing towns of Stirling and Alloa, including brewery and distillery waste. The later development of the port and industries at Grangemouth further down river, added to this cocktail of contamination.

For the first half of the twentieth century, the waters of the upper Forth suffered from a chronic lack of oxygen due to organic and industrial waste, resulting in a much-diminished invertebrate fauna, dominated by pollution-tolerant worm species. It was only with the establishment of the Forth River Purification Board in 1951 that matters began to improve, with other types of invertebrates such as ragworms and small molluscs being recorded by biologists in increasing numbers in the decades that followed. Nowadays, herring and sprat regularly reach as far upstream as Alloa, and most dramatically, a silvery fish known as the sparling has returned to the upper estuary after an absence of many years.

Sparling are fascinating fish, and I recall vividly discovering a shoal of them many years ago at the mouth of the River Almond at Cramond by Edinburgh, where they flickered and flashed in a shallow margin. Sparlings were widely eaten in the past and have a distinctive cucumber aroma when first caught. They are small fish, their slim and slightly translucent bodies growing up to about eight inches – indeed the Scottish name sparling derives from an old French word meaning 'small fish'.

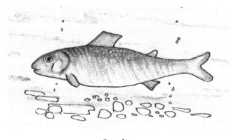

Sparling

On high spring tides in March or early April, sparling make annual spawning influxes into the lower reaches of rivers to spawn at the dead of night. They may spend only a couple of hours in their chosen river before quickly withdrawing to their coastal feeding grounds. Could sparling spawn in the lower reaches of the Devon? It was an exciting possibility, and while the weir on the uppermost part of the estuary would provide a hurdle, it was one that could perhaps be overcome on a high spring tide. Sparling are scarce fish in Scotland that have declined in recent times, currently only recorded in the Firths of Forth and Tay in the east, and in the Solway Firth in the south west.

I soon cast such estuarine reflections aside, and made my way down to the shore, carefully picking a route over slippery rocks and the gloopy thick mud. This area of stone and rocks was in fact made largely of old bricks and other stonework, which I presumed had originated from nearby seawall constructions.

I turned over a stone to reveal a crawling mass of amphipods (small shrimp-like crust-aceans). I was surprised by their abundance, and every rock I turned held the same large numbers beneath. They were larger than the freshwater 'shrimps' found in the main course of the river and were most probably an amphipod species known as the brackish-water shrimp.

Amphipod

The estuary was richer in invertebrates than I had imagined, and these greyish, translucent crustaceans – some almost half-an-inch long – were undoubtedly what the shelducks (and the teal in winter) were feeding upon. These amphipods would also be eagerly devoured by flounders and sea trout, and as such, represented a fundamental keystone in the food chain that helps support the cormorants, goosanders, and seals on the estuary.

Some of the undersurfaces of the rocks also held tiny thread-like reddish worms clinging onto them, which were barely discernible to the eye. I examined one under my magnifying lens, but it was so small that it was hard to discern any detail. It was possibly an oligochaete worm of some type, although it was difficult to be sure.

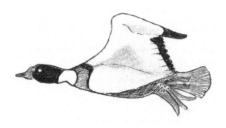

Shelduck

On turning over another rock, a cluster of horse leeches was revealed, a writhing mass of dark bodies. I watched fascinated as one crawled along a rock, the front part of its body elongating, so that the whole creature reached a length of about four inches, before receding again so that the animal became fatter and more rotund. Despite the name, horse leeches do not suck mammalian blood, but feed by ingesting small invertebrates, carrion and dying fish, and sucking the bodily fluids from molluscs. They are freshwater creatures, and their presence at the mouth of the Devon estuary highlighted just how low the salinity of the river was here, despite the tidal influences. There was no denying that these leeches were unappealing to the eye, but it takes all sorts to make the natural world go around, and their place in the ecology of the river was as crucial as any other creature. The only other invertebrate I found were several small empty mollusc shells, which I identified as Hydrobia, a genus of water snails.

I had brought a trowel with me and dug into the mud to see what other creatures lay below. The substrate, however, was thick and black, and as far as I could see, completely devoid of life. Such was the mud's heavy consistency, I doubted many animals could thrive in such an environment, especially since oxygen levels would be extremely low within the tightly compressed sediment. This would also explain the paucity of long-billed waders on the estuary, such as curlews and oystercatchers.

My foray onto the estuary shore had provided an intriguing insight into the invertebrate life found there. While the amphipods provided volume abundance, the estuary, as I had suspected, was much less diverse in life compared to several miles further downstream towards the outer Firth of Forth.

As a boy growing up in Edinburgh in the 1970s, I frequently haunted the shore at Wardie Bay, between Leith and Granton, to search for rockpool fish and other creatures. There was always an Aladdin's Cave of life when turning over rocks, including eelpouts, butterfish, rocklings and short-spined sea scorpion fish. Shore crabs, mussels, limpets, winkles, and whelks were also abundant. I recall the Forth then being heavily polluted with human waste, with there being a sewage outfall just 200 yards offshore. Despite this, I would happily guddle about in these pools brimming with used toilet paper and other unsavoury detritus, completely oblivious to the health risks involved.

Ironically, at that time vast numbers of sea ducks such as scaup, golden-eyes, scoters and eiders congregated near such sewer outfalls where they fed on invertebrate species benefiting from the sewage and from grain discharged from distilleries and breweries, most particularly at Seafield between Leith and Portobello.

A few days later, I returned to the Devon estuary to snorkel – the final fling of my river year. I opted to do so at dawn as the track by the estuary is often frequented by walkers during the day, and I had no wish to attract a curious crowd of onlookers perplexed by the sight of a snorkeller. It was a marvellous early June morning, the air still and the tide at its lowest point, but I was apprehensive as there was a short stretch of mud to wade through before reaching the water, and there was a real danger I might sink deep into it and become irretrievably stuck. To ensure agility, I decided not to wear my flippers and would instead rely on crawling along the shallow bottom of the river or by using swimming actions with my arms and legs for manoeuvring.

I waded tentatively out, but the mud was not as soft as feared, and while it pulled at my wetsuit boots, they sank no deeper than up to my ankles. I reached an algae-covered rock shelf, which I sat upon briefly to don my face mask and snorkel, and then plunged in. With the protection of my wetsuit, the water was surprisingly warm, but also very shallow, and at first, I could not see anything at all as my body and legs stirred up the muddy sediment which swirled around me in a brown cloud. It was a basic mistake. I needed to swim upstream so that river flow carried the detritus away from me, rather than downstream where it went with me. I let myself drift down to a lower rock shelf and then waited for a while for the water to settle slightly, before making my way back upriver where hopefully the visibility would improve. It was certainly clearer, but still extremely murky and I could only see a few inches ahead of me, but that was all I required.

Now relaxed and in a better position, I began to search for creatures. The water was so shallow, I crawled rather than swam up the river, turning over stones with my hands as I went. As was the case with the outermost part of the estuary, there were numerous brackish-water amphipods, which jerkily side-swam through the water every time I flipped a stone over. These amphipods really were the lifeblood of this part of the river, the food that sustains so much else.

Then, the tiniest movement on the muddy bottom, so subtle I could easily have missed it. I looked closer and realised it was a diminutive flounder, no bigger than my thumbnail. I turned a rock, and out scooted another small flounder, and shortly afterwards I glimpsed several more. They would appear briefly, then bury themselves in the mud, disappearing as fast as the flick of a switch.

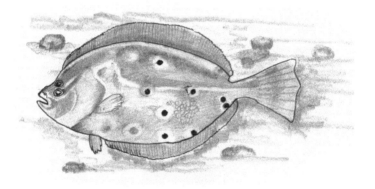

Flounder

I was thrilled. I knew there would be flounders here, but their abundance surprised me, and their tiny size underlined just what an important area the Devon estuary is for these flatfish. In Scotland, adult flounders are estuarine and inshore fish, moving in winter to deeper water in the sea to breed. The previous summer, when snorkelling close to the shore at Kingsbarns in the East Neuk of Fife, I had discovered several adult flounders, about the size of a dinner plate, lying on the seabed, their skin colour matching their surroundings perfectly.

I suspect that some of these tiny juvenile flounders I had found, once they had grown a bit larger, would enter the main river and move upstream to live for a year or so in freshwater, before returning to the estuary and the sea. These flounders were an interesting discovery, and their presence highlighted the environmental importance of estuaries in the wider scale of things as places where young fish can shelter and prosper. Just as how mangrove swamps in the tropics can form important nursery areas for sea fish, then so too do our estuaries.

After a while, rather than snorkelling, I found it more productive to look for life by wading and turning over stones. Many of the undersides of the rocks had little globules of jelly attached to them, which I presumed were amphipod eggs. I also discovered several Hydrobia, as well as cased caddisfly larvae.

I dipped back into the water and floated downstream a short distance to a shallow and exposed stony area, which I trod gingerly down until reaching the narrow muddy channel of the main flow of the last stretch of the Devon that headed out to the Forth. The temptation proved too much, and while I had no chance of seeing anything in the chocolate-stained water here, I slipped back into the water and let the current carry me down the channel towards the Forth, kicking my legs every so often to retain course.

It was a wonderful feeling to sweep down this channel, passing imposing mudbanks on either side, and putting to flight a pair of surprised mallards in the process. Soon, I was at the mouth of the estuary and bobbing about like a cork by the edge of the Forth, where a heron sat nearby on a stony outcrop. I kicked towards the shore and tentatively made my way out of the water by wading through the heavy bankside mud and clambering back up towards the main track, dripping brown ooze all the while.

I pulled off my wetsuit hood and hunkered down onto a grassy knoll to look out over the Forth, both exhilarated and sad at the same time. The estuary had revealed more of its secrets and to explore such a special place was enthralling, especially to experience the pull of the water's flow on that last leg out towards the sea, glimpsing scuttling amphipods and tiny flounders in the process.

My river year had come to a finale, which had left me in a rather melancholy mood. The Devon had become part of my soul and when swimming within its bounds, it was almost as if I had merged into its inner being, feeling its pulse and wonderful embrace.

Chapter 24
RIVER REFLECTIONS
(June)

Early June air, so warm and fair, and my river year had come to an end. I wandered down to my favourite stretch of the middle river and sat on the bankside.

Nature's whisper and colour were all around: the scratchy utterance of a whitethroat from a nearby bramble tangle and a sedge warbler reeling his song from low down in among willows by the water's edge, while the lilac flowers of meadow cranesbill adorned the path edge. The feather-dusted white blooms of meadowsweet were on the verge of emerging; butterflies danced, and bees buzzed. It was almost as if the riverside was singing.

Meadowsweet

The previous few weeks of spring on the Devon had been enjoyable and full of enlightening encounters: the estuary had revealed a multitude of wildlife surprises, and on other parts of the river I had seen beavers on two further occasions. There were other highlights, too, including watching a mother goosander give a piggyback to one of her youngsters as she swam upriver.

Reflections in the water and reflections in the mind. Had I learned more about the Devon and what made it tick? Yes, but only to a degree, because the more I explored, examined and questioned, the more I realised there was so much still to learn.

I had certainly developed a deeper understanding of the river's nuances and subtleties, but many more secrets lay hidden within its watery hold. I had not even begun to consider water chemistry, or river geology, and had only scratched at the surface of the complex ecological interactions between invertebrates, algae and the multitude of other smaller lifeforms that live within the river's bounds. But that did not really matter, because my 12-month dip into the Devon had delivered a lifetime of memories and reinforced, to an almost unimaginable degree, my deeply held conviction about the importance of our rivers to our environment, and thus to humankind.

The River Devon is special – as are all rivers. From that early trickling on the high plateau of Alva Moss to the tumbling flow through gorges and woodland, and then the final languid meanderings to the estuary, life abounded at every turn and in all shapes, sizes and forms. Dippers, bats, mayflies, water crowfoot, goosanders, trout and minnows – the list is endless. Without rivers our lives would be much impoverished. This wonderful diversity of wildlife is a complex ecological web, with each species depending upon one another, either directly or indirectly.

Male and Female Goosander

Rivers provide continuous ribbons of connectivity, linking different habitats and areas through their wild corridors. Their banks are rich in alders, willows and other trees, and the steep-sided valleys created by their multitude of tributaries enable natural woodland to prosper, supporting so much other natural life. River flood meadows are rich in wildflowers, butterflies and other insects, and the associated natural ponds and oxbow lakes are a haven for amphibians. Rivers are the beating heart of our environment.

Rivers link villages and towns that have developed along them in the past, providing water supply and power to support economic development, enabling communities to prosper and thrive. In some parts of the world, navigable rivers play a vital role in the transportation of goods and people. They drain our landscapes and prevent flooding.

Rivers and their environs are also places for recreation and for people to relax – serene slivers of tranquillity. Whether it be anglers or canoeists, birdwatchers or walkers, rivers have something for everyone. In an age where stress and mental health are major issues affecting society, the calming influence of a flowing river can relax the mind and soothe the soul like nothing else.

Despite this, we have abused our rivers for hundreds of years. In the past, they were open sewers for human and industrial waste; dams and weirs were installed, and their courses altered. Some of that legacy still impacts today, and while our rivers are the cleanest they have been for a considerable time, pollution, albeit in mainly more subtle forms, is an ever-present danger.

Other threats loom in the future, most notably climate change. More extreme weather events, such as storms and heavy rainfall, will result in increased flooding and riverside erosion. Warmer summers will lead to increased water extraction and diminished water flows, affecting river wildlife. Some creatures and plants will adapt, many others will not. Thus, rivers need our protection in a more indirect sense, by working on the wider environmental scale to reduce carbon emissions, and change our lifestyles, including our reliance on throwaway plastics.

I also worry about the declining interest in angling, with fewer young people now participating in the hobby. Angling is a wonderful way for connecting with rivers, and angling associations often play a fundamental role in looking after and monitoring our rivers.

While government and its quangos must lead the way when it comes to strategic planning to protect and enhance our river catchments, grassroots 'people power' is an irresistible driving force, and the more communities can become involved with their local rivers, then the greater the chance they will prosper into the future. To that end, rivers need to be made more accessible to communities, for in many instances, riverbanks are remote and hard to reach. Some excellent work is already ongoing to address this, and a co-ordinated national programme to create river path networks would deliver immense benefits.

A few days before the end of my year-long river journey, I hiked a high circuit above the Glen of Sorrow in the Ochils, the place where the Burn of Sorrow cuts along its base and where those unique isolated trout I had found all those months ago lived. It was a wonderful warm late spring day, not dissimilar to my first journey to the river's source the previous year. Meadow pipits swept up before my feet and skylarks hung high in the air, cascading down their rich, sweet songs. At one stage on the walk, I heard the high-pitched piercing call of a golden plover, although the bird remained hidden in the vastness of the landscape.

My route took in the tops of Whitewisp, Tarmangie, Skythorn and King's Seat hills, offering a panoramic perspective of the course of the Devon, and from where I could glimpse the Upper Glendevon Reservoir and the high top of Ben Cleuch, behind which lay Alva Moss and the river's source.

From the summit of King's Seat, a wonderful vista unveiled before me to the south and east, including the gentle meanders of the Devon as it flowed along the foot of the steep Ochil scarp and out into the inner Forth at Cambus.

I sighed in contentment, for it was an inspiring view. In the far distance, I could just discern the Bass Rock in the outer Firth of Forth surrounded by a shimmering, silvery sea. A thought passed through my mind: perhaps, at this very moment, some molecules of water that had spilled out from the mouth of the Devon, many miles further downstream in the inner Forth, were swirling around the Bass Rock. If so, some of this Devon water will inevitably evaporate into the atmosphere and fall once more as rain, perhaps on the high mountain plateaux of Norway or as far away as Russia or beyond.

These raindrops will seep into the ground and develop into trickling rivulets of water, joining streams and then turning into mighty rivers. The truth is there is no real end to any river's story, for each ending turns seamlessly into a new beginning. *If rivers could sing*, what incredible stories their songs would tell.

ABOUT THE AUTHOR

With an interest in wildlife spanning from his earliest years, Keith Broomfield is a well-known Scottish nature writer with a passion for the great outdoors. Keith blogs at http://www.keithbroomfield.com and has a Twitter presence @BroomfieldKeith.

A graduate in zoology from the University of Aberdeen, Keith's writing covers virtually every element of the natural world from flora and fungi, to invertebrates, mammals, birds and marine life.

Based in Strathdevon, Keith gains much of his inspiration from exploring the hills, woods, rivers and lochs of this beautiful part of Scotland. He currently writes a weekly 'Nature Watch' column for *The Courier* newspaper, as well as his 'On the Wildside' column for the *Alloa Advertiser*.

Keith is a trustee of the Forth Rivers Trust and on the committee of the Devon Angling Association.

THE PUBLISHER

Tippermuir Books Ltd (*est.* 2009) is an independent publishing company based in Perth, Scotland.

OTHER TITLES FROM TIPPERMUIR BOOKS

Spanish Thermopylae (2009)

Battleground Perthshire (2009)

Perth: Street by Street (2012)

Born in Perthshire (2012)

In Spain with Orwell (2013)

Trust (2014)

Perth: As Others Saw Us (2014)

Love All (2015)

A Chocolate Soldier (2016)

The Early Photographers of Perthshire (2016)

Taking Detective Stories Seriously:
The Collected Crime Reviews of Dorothy L. Sayers (2017)

Walking with Ghosts (2017)

No Fair City: Dark Tales from Perth's Past (2017)

The Tale o the Wee Mowdie that wantit tae ken
wha keeched on his heid (2017)

Hunters: Wee Stories from the Crescent.
A Reminiscence of Perth's Hunter Crescent (2017)

Flipstones (2018)

Perth: Scott's Fair City: The Fair Maid of Perth & Sir Walter Scott –
A Celebration & Guided Tour (2018)

God, Hitler and Lord Peter Wimsey: Selected Essays,
Speeches and Articles by Dorothy L. Sayers (2019)

Perth & Kinross: A Pocket Miscellany:
A Companion for Visitors and Residents (2019)

The Piper of Tobruk – Pipe Major Robert Roy, MBE, DCM (2019)

The 'Gig Docter o Athole':
Dr William Irvine & The Irvine Memorial Hospital (2019)

Afore the Highlands: The Jacobites in Perth, 1715-16 (2019)

Authentic Democracy: An Ethical Justification of Anarchism (2020)

A Squatter o Bairnrhymes (2020)

FORTHCOMING

Perth Riverside Nurseries: A Spirit of Enterprise and Improvement
(Elspeth Bruce and Pat Kerr, 2020)

The Black Watch in the First World War, 1914-18
(Fraser Brown and Derek Patrick (eds)), 2020)

The Nicht Afore Christmas (Irene McFarlane, 2020)

'The Polis': Darkest Days of the Scottish Police Forces (Gary Knight, 2020/1)

William Soutar: Collected Poetry
(Kirsteen McCue and Paul S Philippou (eds), 2021)

BY LULLABY PRESS
(AN IMPRINT OF TIPPERMUIR BOOKS)
A Little Book of Carol's (2018)

Diverted Traffic (2020)

All Tippermuir Books titles are available from bookshops and online
booksellers. They can also be purchased directly (with free postage &
packing (UK only) – minimum charges for overseas delivery) from

www.tippermuirbooks.co.uk.

Tippermuir Books Ltd can be contacted at
mail@tippermuirbooks.co.uk.